Memoirs of the American Mathematical Society
Number 238

Yum-Tong Siu and
Günther Trautmann

Deformations of coherent analytic sheaves with compact supports

Published by the
AMERICAN MATHEMATICAL SOCIETY
Providence, Rhode Island, USA

January 1981 · Volume 29 · Number 238 (first of 2 numbers)

ABSTRACT

This paper is devoted to the construction of a semi-universal deformation of any coherent analytic sheaf on a complex space which has compact support. The procedure is constructive and elementary. It uses the power series method and the division and extension theory of ideals in power series rings developed by Grauert. This procedure has a number of important special features involving new techniques. Some of these features are: (i) the construction of a partial sheaf system resolution where maps for set-inclusions have directions opposite to those in a sheaf system resolution used for the proof of the direct image theorem, (ii) the use of partially privileged boundaries and Hörmander's L^2 estimates of $\bar{\partial}$ to avoid the need of secondary smoothing, (iii) the establishment of bounds independent of e for extending resolutions of kernels of sheaf-homomorphisms from the eth to the $(e + 1)^{st}$ step and for lifting sections through sheaf-homomorphisms and expressing cocycles as coboundaries at the eth step.

AMS (MOS) subject classifications (1970): 32G05 and 32C35.

Library of Congress Cataloging in Publication Data

Siu, Yum-Tong, 1943-
Deformations of coherent analytic sheaves with compact supports.

(Memoirs of the American Mathematical Society ; no. 238 ISSN 0065-9266)
Bibliography: p.
1. Analytic sheaves, Deformation of.
I. Trautmann, Günther, joint author. II. Title.
III. Series: American Mathematical Society.
Memoirs ; no. 238.
QA3.A57 no. 238 [QA331] 510s [514'.224]
ISBN 0-8218-2238-1 80-26105

TABLE OF CONTENTS

DEFORMATIONS OF COHERENT ANALYTIC SHEAVES WITH COMPACT SUPPORTS[1]

BY

YUM-TONG SIU AND GÜNTHER TRAUTMANN

INTRODUCTION

The goal of this paper is to construct a semi-universal deformation of any coherent analytic sheaf on a complex space which has compact support. Our construction uses the power series method and the division and extension theory of ideals in power series rings developed by Grauert [11]. The method we use is constructive and elementary.

The division and extension theory of ideals was first introduced by Grauert to construct semi-universal deformations of isolated singularities of complex-analytic sets [11]. Grauert later used it together with the Banach-analytic space method of Douady [3] to construct semi-universal deformations of compact complex spaces [12]. In [6,7,7a] Forster and Knorr used the more constructive and explicit power series method to replace the Banach-analytic space method in the construction of semi-universal deformations for holomorphic vector bundles over compact complex spaces, for certain types of compact complex manifolds, and for compact complex spaces.

Though the power series method and the division and extension theory of ideals were already employed in the papers cited above, yet the method of construction given in this paper has a number of important special features involving new techniques

---

[1]During the research for this paper the first author was partially supported by a NSF grant and a Sloan Fellowship and the second author was partially supported by a NSF grant and a travel grant from DFG.

Received by the editors July 23, 1979.

not found in these previous papers. Some of these features are: (i) the construction of a partial sheaf system resolution where maps for set-inclusions have directions opposite to those in a sheaf system resolution used for the proof of the direct image theorem (see § 2), (ii) the use of partially privileged boundaries and Hörmander's L^2 estimates of $\bar{\partial}$ to avoid the need of secondary smoothing (see (7.5) and (10.1)), (iii) the establishment of bounds independent of e for extending resolutions of kernels of sheaf-homomorphisms from the eth to the $(e+1)^{st}$ step and for lifting sections through sheaf-homomorphisms and expressing cocycles as coboundaries at the eth step (see § 7 and § 8).

We now sketch very briefly our construction. Suppose $\mathcal{F}_o$ is a coherent analytic sheaf on a complex space X with compact support. Let $\mathcal{U} = \{U_i\}$ be a Stein cover of X. We say that $\sigma_i : \mathcal{O}_X^{q_i} \to \mathcal{O}_X^{p_i}$ on U_i is a <u>defining sheaf-homomorphism</u> for $\mathcal{F}_o$ on U_i if $\mathcal{F}_o|U_i$ is isomorphic to Coker σ_i. We say that $\tau_{ji} : \mathcal{O}_X^{p_i} \to \mathcal{O}_X^{p_j}$ on $U_i \cap U_j$ is a <u>relating sheaf-homomorphism</u> for σ_i and σ_j if τ_{ji} induces a sheaf-homomorphism $\overline{\tau_{ji}}$: Coker $\sigma_i \to$ Coker σ_j on $U_i \cap U_j$ which is an isomorphism. We describe a sheaf $\mathcal{F}$ by such defining sheaf-homomorphisms $\{\sigma_i\}$ and relating sheaf-homomorphisms $\{\tau_{ji}\}$ which satisfy the compatibility condition $\overline{\tau_{ij}}\ \overline{\tau_{jk}} = \overline{\tau_{ik}}$ on $U_i \cap U_j \cap U_k$.

One difficulty with the power series method for the construction of a semi-universal deformation is the convergence problem which necessitates shrinking and enlarging the covering $\mathfrak{U}$. One has to find a way to pass from a smaller covering to a bigger covering, a process which Forster and Knorr dubbed "smoothing", presumably because entities defined on a bigger covering are smoother than those defined on smaller coverings. In our case convergence and smoothing are handled in the following way. We start out with a coherent analytic sheaf $\mathcal{F}_o$ on X with compact support. Since $\mathrm{Ext}^1(X,\mathcal{F}_o,\mathcal{F}_o)$, which is the space of deformations of $\mathcal{F}_o$ over double points, is finite-dimensional over $\mathbb{C}$, we know from the theory of formal deformations developed by Schlessinger [16] and Schuster [17,18] that we have a semi-universal formal deformation. From the formal deformation we construct a convergent deformation step by step (that is, modulo $\mathfrak{m}^{e+1}$, where $\mathfrak{m}$ is the maximal ideal of $\mathbb{C}^m$ at 0 and m is the dimension of the space of deformations over double points). Suppose at step e the deformation $\mathcal{F}_e$ is represented by defining sheaf-homomorphisms $\{\sigma_i^e\}$ and relating sheaf-homomorphisms $\{\tau_{ij}^e\}$ for the covering $\mathfrak{U} = \{U_i\}$. (This is a simplification of the actual procedure used in the construction in this paper. In the actual construction, for technical reasons, we assume that each U_i is privileged for $\mathcal{F}_o$ and $\mathcal{O}_X$. Moreover, instead of the relating sheaf-homomorphisms $\{\tau_{ij}^e\}$, we use relating sheaf-homomorphisms $\tau_{\mu ij}^e$ for σ_i^e and σ_j^e on D_μ for $D_\mu \subset U_i \cap U_j$, where $\{D_\mu\}$ is a finite collection of privileged sets (see (10.1)) such that each $U_i \cap U_j \cap \mathrm{Supp}\,\mathcal{F}_o$ is contained in the union of the following two sets: (i) the union of all D_μ contained in $U_i \cap U_j$, (ii) the union of all $D_\mu \cap D_\nu$ such that for some k, $D_\mu \subset U_i \cap U_k$ and $D_\nu \subset U_j \cap U_k$.) Take a Stein covering $\tilde{\mathfrak{U}} = \{\tilde{U}_i\}$

with $U_i \subset\subset \tilde{U}_i$. We now use the amalgamation of syzygies to construct defining sheaf-homomorphisms $\tilde{\sigma}_i^e$ for $\mathcal{F}_e$ on $\tilde{U}_i$ and relating sheaf-homomorphisms $\tilde{\tau}_{ji}^e$ for $\tilde{\sigma}_i^e$ and $\tilde{\sigma}_j^e$ on $\tilde{U}_i \cap \tilde{U}_j$ (see § 4 and § 10). We then use $\{\tilde{\sigma}_i^e\}$ and $\{\tilde{\tau}_{ji}^e\}$ to construct a partial sheaf system resolution for $\mathcal{F}_e$ with respect to the covering $\tilde{\mathfrak{U}}$ (see § 2). This partial sheaf system resolution is opposite to the usual one used in proving direct image theorems as far as the homomorphisms corresponding to set-inclusions are concerned, because the resolution is used here to calculate the cohomology groups of $\mathcal{H}om(\mathcal{F}_e, \mathcal{F}_e)$ instead of $\mathcal{F}_e$. By using the partial sheaf system, the algebra involved in constructing $\mathcal{F}_{e+1}$ from $\mathcal{F}_e$ is greatly simplified and becomes manageable (see § 3). The new sheaf $\mathcal{F}_{e+1}$ is constructed from the partial sheaf system and we obtain defining sheaf-homomorphisms φ_i^{e+1} for $\mathcal{F}_{e+1}$ on U_i' and relating sheaf-homomorphisms ψ_{ji}^{e+1} for φ_i^{e+1} and φ_j^{e+1} on $U_i' \cap U_j'$, where $U_i \subset\subset U_i' \subset\subset \tilde{U}_i$. At this stage we do not yet have an extension of the partial sheaf system to the $(e+1)^{st}$ step. We now construct $\{\sigma_i^{e+1}\}$ and $\{\tau_{\mu ji}^{e+1}\}$ from $\{\varphi_i^{e+1}\}$ and $\{\psi_{ji}^{e+1}\}$ so that σ_i^{e+1} extends σ_i^e and $\tau_{\mu ji}^{e+1}$ extends $\tau_{\mu ji}^e$. This completes the induction process of going from the eth step to the $(e+1)^{st}$ step.

In keeping track of the bounds, we need Lemmas of the following type (see § 7 and § 8). Let $U' \subset\subset U$ be Stein open subsets of X, $\mathfrak{U}$ (respectively $\mathfrak{U}'$) be a finite Stein cover of G (respectively G') with members of $\mathfrak{U}'$ relatively compact in members of $\mathfrak{U}$, S_e be the space defined by the ideal of the formal deformation of the eth

step, e_o be the step where the Grauert-Weierstrass families for the ideals of the formal deformation stabilize, and $\varphi_{e_o}: \mathcal{O}^q_{U\times S_{e_o}} \to \mathcal{O}^p_{U\times S_{e_o}}$ be a sheaf-homomorphism with S_{e_o}-flat cokernel. Suppose $g_{e_o} \in \Gamma(U\times S_{e_o}, \mathcal{O}^q_{U\times S_{e_o}})$, $f_{e_o} = \varphi_{e_o} g_{e_o}$, $\eta_{e_o} \in C^{s-1}(\mathfrak{U}\times S_{e_o}, \operatorname{Im}\varphi_{e_o})$, and $\xi_{e_o} = \delta\eta_{e_o}$. Let $\rho_o \in \mathbb{R}^m_+$. Then there exist positive functions $L_*(L_o)$, $\varepsilon_*(L,L_o)$, and $\theta_*(L_o,L,\varepsilon)$ such that the following holds. Suppose $e \geq e_o$ and $\varphi_e: \mathcal{O}^q_{U\times S_e} \to \mathcal{O}^p_{U\times S_e}$ with S_e-flat cokernel (respectively $f_e \in \Gamma(U\times S_e, \operatorname{Im}\varphi_e)$, $\xi_e \in Z^s(\mathfrak{U}\times S_e, \operatorname{Im}\varphi_e)$) extend φ_{e_o} (respectively f_{e_o}, ξ_{e_o}). If $L_o > 0$, $0 < L < L_*(L_o)$, $0 < \varepsilon < \varepsilon_*(L_o, L)$, and $0 < \theta < \theta_*(L_o, L, \varepsilon)$ and if $\rho = \theta\rho_o$ and

$$\| \varphi_e - \varphi_{e_o} \|_{U,\rho} \leq L_o \varepsilon \theta^{e_o}$$

$$\| f_e - f_{e_o} \|_{U,\rho} \leq L_o \varepsilon \theta^{e_o}$$

$$\| \xi_e - \xi_{e_o} \|_{\mathfrak{U},\rho} \leq L_o \varepsilon \theta^{e_o},$$

then there exists $g_e \in \Gamma(U'\times S_e, \mathcal{O}^q_{U\times S_e})$ extending g_{e_o} with $\varphi_e g_e = f_e$ and there exists $\eta_e \in C^{s-1}(\mathfrak{U}'\times S_e, \operatorname{Im}\varphi_e)$ extending η_{e_o} with $\delta\eta_e = \xi_e$ such that

$$\| g_e - g_{e_o} \|_{U',\rho} \leq L\varepsilon\theta^{e_o}$$

$$\| \eta_e - \eta_{e_o} \|_{\mathfrak{U}',\rho} \leq L\varepsilon\theta^{e_o},$$

where the semi-norms $\|\cdot\|_{U,\rho}$, $\|\cdot\|_{\mathfrak{U},\rho}$ carry essentially the usual meanings and are precisely defined in § 6. This type of Lemmas is important, because they give a priori estimates independent of e and independent of the knowledge of φ_e.

We use privileged U_i and D_μ to avoid the need of "secondary smoothing". To define privilegedness, we use L^2 norms instead of sup norms. The reason is that we have to consider sets which are only partially privileged. These sets arise when we consider the Čech cohomology with respect to $\{H \cap W_i\}$, where H is privileged and $\{W_i\}$ is a collection of Stein sets covering H^-, in order to piece together sheaf-homomorphisms defined on $H \cap W_i$ to form a sheaf-homomorphism defined on H. To handle such partially privileged sets $H \cap W_i$, we have to show that a $\bar{\partial}$-closed (0,q)-form on $H \cap W_i$ with bounds along the boundary $W_i \cap \partial H$ is the $\bar{\partial}$-image of a (0,q-1)-form on $H \cap W_i$ with bounds along the boundary $W_i \cap \partial H$ (that is, the forms are bounded on $V \cap H$ for any $V \subset\subset W_i$). When we use uniform bounds, there is no such result, but, when we use L^2 bounds we can use Hörmander's L^2 estimates of $\bar{\partial}$ by employing the weight function $e^{-\theta}$, where θ is a suitable plurisubharmonic exhaustion function on W_i chosen differently for each $\bar{\partial}$-closed form.

Our construction of a semi-universal deformation of a coherent analytic sheaf with compact support, together with the existence of a semi-universal deformation of a compact complex space proved by Grauert [12], yields the existence of a semi-universal joint deformation of a compact complex space and a coherent analytic sheaf on it (see § 1).

The construction of morphisms of deformations given in § 11 can be used to decide whether a given deformation is semi-universal.

We would like to remark that Forster-Knorr's version of Cartan's Lemma (cf. (6.11), (6.12)) is used in our construction.

Our method of construction contains two types of arguments: one purely algebraic and another involving estimates. The construction becomes more apparent if, before following the arguments on estimates, one first reads § 2, § 3, § 10 (ignoring the estimates) to get an understanding of the algebraic arguments.

§ 1 Deformations

(1.0) Notations: If $p: X \to S$ is a morphism of complex spaces and if $\mathcal{G}$ is a coherent analytic sheaf on X we denote by $\mathcal{G}(s)$ the analytic restriction of $\mathcal{G}$ to the fibre $X(s) = p^{-1}(s)$. If X is a product $X_o \times S$ and p the caconical projection, $\mathcal{G}(s)$ will be considered as a sheaf on X_o, i.e. $\mathcal{G}(s)$ is the sheaf $j_s^*\mathcal{G}$ where $j_s: X_o \to X_o \times S$ is the canonical injection $x \mapsto (x,s)$. Any coherent sheaf $\mathcal{G}$ on $X_o \times S$ thus defines a family $\{\mathcal{G}(s)\}_{s \in S}$ of sheaves on X_o. The sheaf $\mathcal{G}$ is called S-flat if for any $x \in X$ the stalk $\mathcal{G}_x$ is flat as a module over $\mathcal{O}_{S,p(x)}$ via the canonical homomorphism $\mathcal{O}_{S,p(x)} \to \mathcal{O}_{X,x}$.

(1.1) Definition: Let $\mathcal{F}$ be a coherent analytic sheaf on the complex space X. A deformation of $\mathcal{F}$ is a quadruple $(\mathcal{G},S,s_o,\alpha)$ where (S,s_o) is the germ of a complex space with representative S and where $\mathcal{G}$ is a coherent S-flat sheaf on $X \times S$, such that $\alpha: \mathcal{G}(s_o) \to \mathcal{F}$ is an isomorphism of $\mathcal{O}_X$-modules on X.

If $(\mathcal{G},S,s_o,\alpha)$ is a deformation and if $f: (S',s_o') \to (S,s_o)$ is a morphism of germs, then $((id \times f)^*\mathcal{G},S',s_o',\alpha')$ with $\alpha' = \alpha: j_{s_o'}^*(id \times f)^*\mathcal{G} = j_{s_o}^*\mathcal{G} \to \mathcal{F}$ is again a deformation because $(id \times f)^*\mathcal{G}$ is S'-flat, [3].

A morphism $(\mathcal{G},S,s_o,\alpha) \to (\mathcal{G}',S',s_o',\alpha')$ of two deformations of $\mathcal{F}$ is defined as a pair (φ,f) where $f: (S',s_o') \to (S,s_o)$ is a morphism of germs and $\varphi: (id \times f)^*\mathcal{G} \to \mathcal{G}'$ is an isomorphism on $X \times S'$ such that $\alpha = \alpha' \circ j_{s_o'}^*(\varphi)$. In this case for any $s' \in S'$

we have $\mathcal{G}'(s') \cong \mathcal{G}(fs')$ and hence $\{\mathcal{G}'(s')\}_{s'\in S'}$ is a subfamily of the family $\{\mathcal{G}(s)\}_{s\in S}$. In the sequal we will omit s_o and α and only write $(\mathcal{G},S)$ instead of $(\mathcal{G},S,s_o,\alpha)$.

(1.2) Definition. The deformation $(\mathcal{G},S)$ is called complete if for any other deformation $(\mathcal{G}',S')$ there exists a morphism $(\varphi,f):(\mathcal{G},S) \to (\mathcal{G}',S')$. If in addition the tangential map $Tf: T_{s_o'}S' \to T_{s_o}S$ is the same for all such morphisms, the deformation $(\mathcal{G},S)$ is called semi-universal or versal.

(1.3) If $(\mathcal{G},S)$ is a versal deformation of $\mathcal{F}$ the space $T_{s_o}S$ agrees with the $\mathbb{C}$-vector space $\mathrm{Ext}^1(X,\mathcal{F},\mathcal{F})$ and this in turn agrees with the space of deformations of $\mathcal{F}$ over the double point, [22]. Especially $\dim_{\mathbb{C}}\mathrm{Ext}^1(X,\mathcal{F},\mathcal{F}) < \infty$ is a necessary condition for the existence of a versal deformation. The converse is not true in general, but if $\dim_{\mathbb{C}}\mathrm{Ext}^1(X,\mathcal{F},\mathcal{F}) < \infty$ there exists a formal versal deformation, [16], [18]. If Supp $(\mathcal{F})$ is compact then $\mathrm{Ext}^1(X,\mathcal{F},\mathcal{F})$ is finite-dimensional as can be easily seen by the spectral sequence defined by $H^i(X,\mathcal{E}xt^j(\mathcal{F},\mathcal{F}))$. The purpose of this paper is to prove the following

Theorem I: *Let X be an arbitrary complex space and let $\mathcal{F}$ be a coherent analytic sheaf on X with compact support. Then $\mathcal{F}$ has a versal deformation.*

(1.4) Joint deformation of a compact complex space and a coherent sheaf. Let X_o be a compact complex space and $\mathcal{F}_o$ be a coherent sheaf on X_o. We define a joint deformation of the pair $(X_o, \mathcal{F}_o)$ as a tuple

$$(X, \pi, S, s_o, \iota, \mathcal{G}, \alpha)$$

where (X, π, S, s_o, ι) is a deformation of X_o as indicated by the diagram $\boxdot$, [12], [5], and where $\mathcal{G}$ is a coherent S-flat sheaf on X together with an isomorphism

$$\begin{array}{ccc} X_o & \overset{\iota}{\hookrightarrow} & X \\ \downarrow & \boxdot & \downarrow{\scriptstyle \pi} \\ \{s_o\} & \hookrightarrow & S \end{array}$$

$$\alpha : \mathcal{G}(s_o) = \iota^*\mathcal{G} \to \mathcal{F}_o .$$

Again we use the abbreviation $(X, S, \mathcal{G})$ for such a deformation. A morphism

$$(X, S, \mathcal{G}) \longrightarrow (X', S', \mathcal{G}')$$

of two joint deformations is a triple $(f, \tilde{f}, \varphi)$ where $(f, \tilde{f})$ defines a morphism of the deformations of spaces, compatible with the ι's, as indicated in the cartesian diagram $\#$ and where

$$\begin{array}{ccc} X' & \overset{\tilde{f}}{\longrightarrow} & X \\ \downarrow{\scriptstyle \pi'} & \# & \downarrow{\scriptstyle \pi} \\ S' & \underset{f}{\longrightarrow} & S \end{array}$$

$$\varphi : \tilde{f}^*\mathcal{G} \to \mathcal{G}'$$

is an isomorphism on X' which is compatible with the α's.

As is (1.2) the joint deformation $(X, S, \mathcal{G})$ is called <u>versal</u>, if there exists a morphism to any other joint deformation with unique tangential map. Using theorem I and the existence of a versal deformation of a compact space proved by Grauert [12], Douady [5] and Forster-Knorr [7a], one can derive the existence of a joint versal deformation in the compact case:

<u>Theorem II: If $\mathcal{F}_o$ is a coherent analytic sheaf on a compact complex space X_o, then the pair $(X_o, \mathcal{F}_o)$ has a joint versal deformation.</u>

(1.5) Proof: Let (X,π,S,s_o,ι) be the versal deformation of X_o with pullback diagram ⊡ as in (1.4). Let at the same time $\pi: X \to S$ be the map over a representative S of the germ (S,s_o). The sheaf $\iota_*\mathcal{F}_o$ still has compact support contained in the fibre X_o. By theorem I there exists a versal deformation $(\mathcal{F},T,t_o,\alpha)$ of $\iota_*\mathcal{F}_o$ with X fixed and $\mathcal{F}$ coherent and T-flat on $X \times T$.

(1.5.1) Remark: $\mathcal{F}$ is at the same time versal in the category of deformations of $\iota_*\mathcal{F}_o$ when X is regarded as a germ with respect to S. For if $S' \subset S$ is a neighborhood of s_o and $X' = \pi^{-1}(S')$ and if $(\mathcal{G}',T',t_o',\alpha')$ is a deformation of $\iota_*\mathcal{F}_o$ living on $X' \times T'$ we have $\mathrm{Supp}(\mathcal{G}') \cap (\pi^{-1}(\partial S'') \times t_o') = \emptyset$ for any neighborhood S'' of s_o relatively compact in S'. Hence by shrinking T' we can extend $\mathcal{G}'|\pi^{-1}(S'') \times T'$ trivially to a sheaf $\mathcal{G}$ on $X \times T'$ which still represents a deformation of $\iota_*\mathcal{F}_o$. Now we obtain a morphism $(f,\varphi) : (\mathcal{F},T) \to (\mathcal{G},T')$ which defines a morphism of the germs with respect to $S \times T$.

(1.5.2) Still regarding S and T as spaces and not as germs, we use a result of Frisch [8] to transform $\mathcal{F}$ into a flat sheaf over a new basis. If $b : B \to S \times T$ is any morphism we denote by $X(B)$ the pullback of $X \times T$ via b and by $\mathcal{F}(B)$ the pullback of $\mathcal{F}$ on $X(B)$. By [8] there exists a complex space R and a bijective immersion $r : R \to S \times T$ such that: (i) $\mathcal{F}(R)$ is R-flat, (ii) if $B \xrightarrow{b} S \times T$ is any morphism such that $\mathcal{F}(B)$ is B-flat then b factors uniquely through r.

(1.5.3) The triple $(X(R),R,\mathcal{F}(R))$ represents the desired joint versal deformation when regarded as over the germ (R,r_o) with $r(r_o) = (s_o,t_o)$. To verify this let $(Y,B,\mathcal{G})$ be any joint deformation. Since (X,S) is the versal deformation of the space X_o, there exists a pair $(f,\tilde{f})$ of morphisms such that the corresponding square $*$ is cartesian. Let $j := (\tilde{f},\pi_Y)$ be the canonical

$$\begin{array}{ccc} Y & \xrightarrow{\tilde{f}} & X \\ {\scriptstyle \pi_Y}\big\downarrow & * & \big\downarrow{\scriptstyle \pi_X} \\ B & \xrightarrow[f]{} & S \end{array} \qquad\qquad \begin{array}{ccc} X_o & \xrightarrow{\iota_X} & X \\ {\scriptstyle \iota_Y}\big\downarrow & \# & \big\downarrow{\scriptstyle i_o} \\ Y & \xrightarrow[j]{} & X\times B \end{array}$$

imbedding $Y \hookrightarrow X\times B$ and let $i_o = (id,b_o)$ where b_o is the distinguished point of B. Then the diagram $\#$ is commutative. We have:

(1.5.4) $j_*\mathcal{G}$ <u>represents a deformation of</u> $\iota_{X*}\mathcal{F}_o$.

To show this, note first that the square $\#$ is cartesian by using the fact that ι_Y is a pullback-map. Hence $i_o^{-1}j(Y)$ identifies with X_o via ι_X and we have $\iota_{X*}\,\iota_X^*\,i_o^*\,j_*\mathcal{G} = i_o^*\,j_*\mathcal{G}$. Since j is an imbedding we also have $j^*j_*\mathcal{G} = \mathcal{G}$.

Now $\iota_Y^*\mathcal{G} \cong \mathcal{F}_o$ implies $i_o^*\,j_*\mathcal{G} = \iota_{X*}\,\iota_X^*\,i_o^*\,j_*\mathcal{G} = \iota_{X*}\,\iota_Y^*\,j^*\,j_*\mathcal{G}$

$= \iota_{X*}\,\iota_Y^*\,\mathcal{G} \cong \iota_{X*}\mathcal{F}_o$. Since $j_*\mathcal{G}$ is also B-flat by $pr\circ j = \pi_Y$ and the definition of flatness, hence (1.5.4) is verified.

(1.5.5) To obtain a morphism $(X(R),R,\mathcal{F}(R)) \to (Y,B,\mathcal{G})$ we define one after another the morphisms in the following diagram

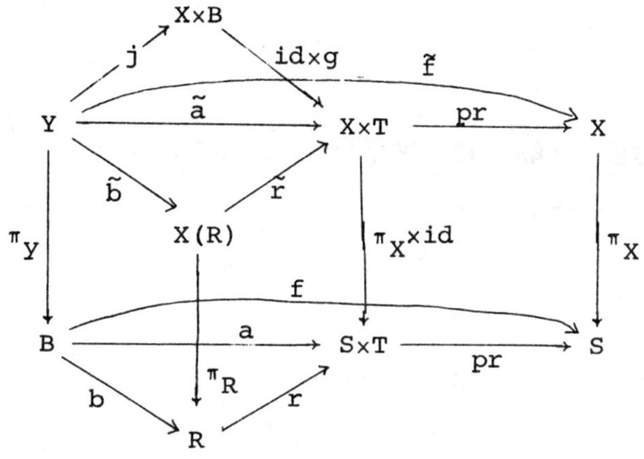

The morphisms π_X, π_Y, π_R are the given space deformations, $\tilde{r}$ is the pullback map over r such that $\mathcal{F}(R) = \tilde{r}^*\mathcal{F}$. The maps $f, \tilde{f}$ and j are defined as in (1.5.3). Since $j_*\mathcal{G}$ is a deformation of $\iota_{X*}\mathcal{F}_o$, by the definition of $(\mathcal{F}, T)$ there exists a morphism $g : B \to T$ such that

$$(id \times g)^*\mathcal{F} = j_*\mathcal{G}.$$

Now we define $a = (f,g)$ and $\tilde{a} = (\tilde{f}, g\ \pi_Y)$. Then $pr\ a = f$, $pr\ \tilde{a} = \tilde{f}$ and moreover

$$(id \times g)\circ j = (\tilde{f}, g\circ\pi_Y) = \tilde{a}$$

$$(\pi_X \times id)\circ\tilde{a} = (\pi_X\circ\tilde{f}, g\circ\pi_Y) = a\circ\pi_Y.$$

Since the squares with $(f,\tilde{f})$ and (pr,pr) are cartesian, the square $(a,\tilde{a})$ is cartesian. Now $\mathcal{G} = j^*j_*\mathcal{G} \cong j^*(id\times g)^*\mathcal{F} = \tilde{a}^*\mathcal{F}$ and hence $\tilde{a}^*\mathcal{F}$ is B-flat. By (1.5.2) the morphism a factorizes uniquely into $a = r\circ b$ and since the square with $(r,\tilde{r})$ is cartesian, there exists a unique morphism $\tilde{b}$ with $\tilde{a} = \tilde{r}\circ\tilde{b}$ and $\pi_R\circ\tilde{b} = b\circ\pi_Y$. Again the square with $(b,\tilde{b})$ is cartesian. Moreover $\mathcal{G} \cong \tilde{a}^*\mathcal{F} = \tilde{b}^*\tilde{r}^*\mathcal{F} = \tilde{b}^*\mathcal{F}(R)$ and there exists an isomorphism $\varphi : \tilde{b}^*\mathcal{F}(R) \longrightarrow \mathcal{G}$. It is easy to check, that this isomorphism respects the canonical isomorphisms with $\mathcal{F}_o$. By $(b,\tilde{b},\varphi)$ we thus obtain a morphism $(X(R),R,\mathcal{F}(R)) \to (Y,B,\mathcal{G})$ of the joint deformations.

(1.5.6) Finally we have to show that the map Tb is unique. For this purpose we show that any morphism $(b,\tilde{b},\varphi) : (X(R),R,\mathcal{F}(R)) \to (Y,B,\mathcal{G})$ arises in the above way by using the same diagram as in (1.5.5) only with a new order of definition of the arrows. If $(b,\tilde{b},\varphi)$ is given with $\varphi : \tilde{b}^*\mathcal{F}(R) \to \mathcal{G}$ an isomorphism, define $\tilde{a} = \tilde{r}\circ\tilde{b}$, $a = r\circ b$, $\tilde{f} = pr_1\circ\tilde{a}$, $f = pr_1\circ a$, $g = pr_2\circ a$ and $j = (\tilde{f},\pi_Y)$. It follows now that $a = (f,g)$ and $(id \times g)\, j = (\tilde{f},g\circ\pi_Y) = (\tilde{f},pr_2\circ a\circ\pi_Y) = (\tilde{f},pr_2\circ(\pi_X \times id)\circ\tilde{a})$ $= (pr_1\circ\tilde{a},pr_2\circ\tilde{a}) = \tilde{a}$. Since by composition the squares with $(a,\tilde{a})$ and $(f,\tilde{f})$ are cartesian, j is the canonical imbedding of $Y \hookrightarrow X \times B$. Moreover Tf is unique, because (X,S) is versal. Since $\mathcal{G} = \tilde{b}^*\mathcal{F}(R) = \tilde{a}^*\mathcal{F} = j^*(id \times g)^*\mathcal{F}$ we obtain an epimorphism $\beta : (id \times g)^*\mathcal{F} \to j_*j^*(id \times g)^*\mathcal{F} \cong j_*\mathcal{G}$. By (1.5.4) $j_*\mathcal{G}$ is a deformation of $\iota_{X*}\mathcal{F}$. Since also $(id \times g)^*\mathcal{F}$ is such a deformation, $Ker(\beta)$ vanishes on X_o and also on $\pi_X^{-1}(S')$ for some neighborhood $S' \subset\subset S$ of s_o. Hence as germs $(id\times g)^*\mathcal{F} = j_*\mathcal{G}$ and since $(\mathcal{F},T)$ was versal, also Tg is unique. Now by $r\circ b = a = (f,g)$ we have $Tr\circ Tb = (Tf,Tg)$. Since Tr is injective, for r is an immersion, Tb is uniquely determined by Tf and Tg, hence uniquely determined by the joint deformation $(Y,B,\mathcal{G})$.

§ 2 Sheaf systems

Let X be a complex space and $(U_i)_{i\in I}$ be a locally finite covering of X by Stein open subsets. Assume that a coherent analytic sheaf $\mathcal{F}$ has a resolution on each U_i as follows:

$$\mathcal{O}^{s_i} \xrightarrow[c_i]{} \mathcal{O}^{r_i} \xrightarrow[b_i]{} \mathcal{O}^{q_i} \xrightarrow[a_i]{} \mathcal{O}^{p_i} \xrightarrow[\pi_i]{} \mathcal{F}|U_i \longrightarrow 0.$$

Since also the intersections $U_{ij} = U_i \cap U_j$ are Stein, by Theorem B it is possible to find homomorphisms φ_{ij}, ψ_{ij}, θ_{ij} over U_{ij} relating the local resolutions as indicated in the following diagram (#).

$$\begin{array}{ccccccccc}
\mathcal{O}^{s_i} & \xrightarrow{c_i} & \mathcal{O}^{r_i} & \xrightarrow{b_i} & \mathcal{O}^{q_i} & \xrightarrow{a_i} & \mathcal{O}^{p_i} & \xrightarrow{\pi_i} & \mathcal{F}|U_{ij} \\
 & & \downarrow{\scriptstyle \theta_{ji}} & & \downarrow{\scriptstyle \psi_{ji}} & & \downarrow{\scriptstyle \varphi_{ji}} & & \| \\
\mathcal{O}^{s_j} & \xrightarrow[c_j]{} & \mathcal{O}^{r_j} & \xrightarrow[b_j]{} & \mathcal{O}^{q_j} & \xrightarrow[a_j]{} & \mathcal{O}^{p_j} & \xrightarrow[\pi_j]{} & \mathcal{F}|U_{ij}
\end{array} \qquad (\#)$$

By the same argument we also obtain homomorphisms f_{kji}, g_{kji}, h_{kji} over U_{ijk} which are homotopies for the above relating homomorphisms, such that we have the following identities on U_{ijk}:

$$\varphi_{ki} = \varphi_{kj}\circ\varphi_{ji} + a_k\circ f_{kji}$$
$$\psi_{ki} = \psi_{kj}\circ\psi_{ji} + f_{kji}\circ a_i + b_k\circ g_{kji}$$
$$\theta_{ki} = \theta_{kj}\circ\theta_{ji} + g_{kji}\circ b_i + c_k\circ h_{kji} .$$

Again by the same arguments and by considering $\varphi_{\ell k}\circ\varphi_{kj}\circ\varphi_{ji}$ and $\psi_{\ell k}\circ\psi_{kj}\circ\psi_{ji}$ on $U_{ijk\ell}$ we obtain homomorphisms $F_{\ell kji}$ and $G_{\ell kji}$ relating the first two homotopies:

$$f_{\ell ki} - f_{\ell ji} = f_{\ell kj}\circ\varphi_{ji} - \psi_{\ell k}\circ f_{kji} + b_\ell\circ F_{\ell kji}$$
$$g_{\ell ki} - g_{\ell ji} = g_{\ell kj}\circ\psi_{ji} - \theta_{\ell k}\circ g_{kji} + c_\ell\circ G_{\ell kji}$$

<u>(2.1) Definition:</u> Let $(U_i)_{i\in I}$ and $\mathcal{F}$ be as above. By I^p we

denote the set of ordered tuples $\alpha = (i_1,\dots,i_p)$ with $i_\nu \neq i_\mu$ for $\nu \neq \mu$ and we write $\alpha \hookrightarrow \beta$ if we have an order-preserving inclusion for $\alpha \in I^p$, $\beta \in I^q$. We write $|\alpha| = p$ if $\alpha \in I^p$ and $U_\alpha = U_{i_1} \cap \dots \cap U_{i_p}$. By I^* we denote the disjoint union of all I^p, $p \geq 1$.

By a (free and inverse) sheaf system with respect to $(U_i)_{i\in I}$ we mean a system

$$\mathcal{R}_* = (\mathcal{R}_\alpha, \Phi_{\alpha\beta})_{\alpha,\beta\in I^*}$$

where for each α, $\mathcal{R}_\alpha$ is a finite free $\mathcal{O}_X$-module and for each pair $\alpha \hookrightarrow \beta$, $\Phi_{\alpha\beta} : \mathcal{R}_\beta \to \mathcal{R}_\alpha | U_\beta$ is a sheaf homomorphism over U_β, such that for $\alpha \hookrightarrow \beta \hookrightarrow \gamma$ we have the identity

$$\Phi_{\alpha\beta} \circ \Phi_{\beta\gamma} = \Phi_{\alpha\gamma} .$$

A homomorphism $A_* : \mathcal{R}_* \to \mathcal{R}'_*$ between two such systems is a system $(A_\alpha)_{\alpha\in I^*}$ of homomorphisms $A_\alpha : \mathcal{R}_\alpha \to \mathcal{R}'_\alpha$ satisfying $\Phi'_{\alpha\beta} A_\beta = A_\alpha \Phi_{\alpha\beta}$.

(2.2) We are going to assign to the system (#) of resolutions of $\mathcal{F}$ part of a resolution

$$\to \mathcal{R}^2_* \to \mathcal{R}^1_* \to \mathcal{R}^o_* \to \mathcal{F}|U_* \to 0$$

of $\mathcal{F}$ by sheaf systems, i.e. to construct sheaf systems and homomorphisms of sheaf systems such that for any $\alpha \in I^*$ we have a resolution of $\mathcal{F}|U_\alpha$. We only construct $\mathcal{R}^p_\alpha$ for $|\alpha|+p \leq 4$, which satisfies our needs:

a) For α = (i), (ij), (ijk), (ijkℓ) the rows

$$\mathcal{R}^3_i \xrightarrow[C_i]{} \mathcal{R}^2_i \xrightarrow[B_i]{} \mathcal{R}^1_i \xrightarrow[A_i]{} \mathcal{R}^o_i \xrightarrow[P_i]{} \mathcal{F}|U_i$$

$$\mathcal{R}^2_{ij} \xrightarrow[B_{ij}]{} \mathcal{R}^1_{ij} \xrightarrow[A_{ij}]{} \mathcal{R}^o_{ij} \xrightarrow[P_{ij}]{} \mathcal{F}|U_{ij}$$

$$\mathcal{R}^1_{ijk} \xrightarrow[A_{ijk}]{} \mathcal{R}^o_{ijk} \xrightarrow[P_{ijk}]{} \mathcal{F}|U_{ijk}$$

$$\mathcal{R}^o_{ijk\ell} \longrightarrow \mathcal{F}|U_{ijk\ell}$$

will be defined respectively by the following rows:

$$\mathcal{O}^{s_i} \xrightarrow[C_i]{} \mathcal{O}^{r_i} \xrightarrow[B_i]{} \mathcal{O}^{q_i} \xrightarrow[A_i]{} \mathcal{O}^{p_i} \xrightarrow[P_i]{} \mathcal{F}|U_i$$

$$\mathcal{O}^{r_i}\oplus\mathcal{O}^{r_j}\oplus\mathcal{O}^{s_j} \xrightarrow[B_{ij}]{} \mathcal{O}^{q_i}\oplus\mathcal{O}^{q_j}\oplus\mathcal{O}^{r_j} \xrightarrow[A_{ij}]{} \mathcal{O}^{p_i}\oplus\mathcal{O}^{q_j} \xrightarrow[P_{ij}]{} \mathcal{F}|U_{ij}$$

$$\mathcal{O}^{q_i}\oplus\mathcal{O}^{q_j}\oplus\mathcal{O}^{q_k}\oplus\mathcal{O}^{r_j}\oplus\mathcal{O}^{r_k} \xrightarrow[A_{ijk}]{} \mathcal{O}^{p_i}\oplus\mathcal{O}^{q_j}\oplus\mathcal{O}^{q_k}\oplus\mathcal{O}^{r_k} \xrightarrow[P_{ijk}]{} \mathcal{F}|U_{ijk}$$

$$\mathcal{O}^{p_i}\oplus\mathcal{O}^{q_j}\oplus\mathcal{O}^{q_k}\oplus\mathcal{O}^{q_\ell}\oplus\mathcal{O}^{r_k}\oplus\mathcal{O}^{r_\ell} \xrightarrow[P_{ijk\ell}]{} \mathcal{F}|U_{ijk\ell},$$

where the homomorphisms are defined as follows:

(we write x_i, y_i, z_i for the elements in $\mathcal{O}^{p_i}$, $\mathcal{O}^{q_i}$, $\mathcal{O}^{r_i}$ respectively)

$P_i x_i = P_{ij}(x_i\oplus y_j) = P_{ijk}(x_i\oplus\ldots) = P_{ijk\ell}(x_i\oplus\ldots) = \pi_i x_i$,

$A_i y_i = a_i y_i$, $A_{ij}(y_i\oplus y_j\oplus z_j) = a_i y_i + y_j$,

$$A_{ijk}(y_i\oplus y_j\oplus y_k\oplus z_j\oplus z_k) = a_i y_i + y_j + y_k + z_k ,$$

$B_i z_i = b_i z_i$, $B_{ij}(z_i\oplus z_j\oplus u_j) = b_i z_i + 0 + z_j$,

$C_i u_i = c_i u_i$.

It is easy to show that with these definitions the above rows are exact.

b) We write $\Phi_{\alpha\beta}$, $\Psi_{\alpha\beta}$, $\Theta_{\alpha\beta}$ for the relating homomorphisms in the sheaf-systems $\mathcal{R}^o_*$, $\mathcal{R}^1_*$, $\mathcal{R}^2_*$ respectively and define these homomorphisms (with obvious notations) as follows.

$$z_i \xleftarrow{\Theta_{i,ij}} z_i \oplus z_j \oplus u_j \xrightarrow{\Theta_{j,ij}} \theta_{ji} z_i \oplus z_j \oplus c_j u_j$$

$$y_i \xleftarrow{\Psi_{i,ij}} y_i \oplus y_j \oplus z_j \xrightarrow{\Psi_{j,ij}} \psi_{ji} y_i \oplus y_j \oplus b_j z_j$$

$$x_i \xleftarrow{\Phi_{i,ij}} x_i \oplus y_i \xrightarrow{\Phi_{j,ij}} \varphi_{ji} x_i \oplus a_j y_j$$

$$y_i \oplus y_j \oplus 0 \xleftarrow{\Psi_{ij,ijk}} y_i \oplus y_j \oplus y_k \oplus z_j \oplus z_k \xrightarrow{\Psi_{ik,ijk}} y_i \oplus (y_k \oplus \psi_{kj} y_j) \oplus z_k$$

$$y_i \oplus y_j \oplus y_k \oplus z_j \oplus z_k \xrightarrow{\Psi_{jk,ijk}} (\psi_{ji} y_i + y_j) \oplus (y_k + b_k z_k + f_{kji} a_i y_i) \oplus (g_{kji} y_i)$$

$$x_i \oplus y_j \xleftarrow{\Phi_{ij,ijk}} x_i \oplus y_j \oplus y_k \oplus z_k \xrightarrow{\Phi_{ik,ijk}} x_i \oplus (y_k + \psi_{kj} y_j)$$

$$x_i \oplus y_j \oplus y_k \oplus z_k \xrightarrow{\Phi_{jk,ijk}} (\varphi_{ji} x_i + a_j y_j) \oplus (y_k + b_k z_k + f_{kji} x_i)$$

and we finally define

$$x_i \oplus 0 \oplus 0 \oplus z_k \xleftarrow{\Phi_{ijk,ijk\ell}} x_i \oplus y_i \oplus y_k \oplus y_\ell \oplus z_k \oplus z_\ell \xrightarrow{\Phi_{ij\ell,ijk\ell}} x_i \oplus 0 \oplus (\psi_{\ell k} f_{kji} x_i) \oplus (\theta_{\ell k} z_k)$$

$$x_i \oplus y_i \oplus y_k \oplus y_\ell \oplus z_k \oplus z_\ell \xrightarrow{\Phi_{jk\ell,ijk\ell}} (\varphi_{ji} x_i) \oplus (f_{kji} x_i + b_k z_k) \oplus (f_{\ell ji} x_i) \oplus (F_{\ell kji} x_i)$$

$$x_i \oplus y_i \oplus y_k \oplus y_\ell \oplus z_k \oplus z_\ell \xrightarrow{\Phi_{ik\ell,ijk\ell}} x_i \oplus 0 \oplus (\psi_{\ell k} f_{kji} x_i) \oplus 0$$

c) Let $|\alpha|=1$, $|\beta|=2$, $|\gamma|=3$, $|\delta|=4$. It is straightforward to verify that $A_\alpha \Psi_{\alpha\beta} = \Phi_{\alpha\beta} A_\beta$, $B_\alpha \Theta_{\alpha\beta} = \Psi_{\alpha\beta} B_\beta$, $A_\beta \Psi_{\beta\gamma} = \Phi_{\alpha\beta} A_\beta$ and that $\Phi_{\alpha\gamma} := \Phi_{\alpha\beta}\Phi_{\beta\gamma}$, $\Phi_{\beta\dot{\gamma}} := \Phi_{\beta\gamma}\Phi_{\gamma\delta}$, $\Phi_{\alpha\delta} := \Phi_{\alpha\beta}\Phi_{\beta\delta}$ and $\Psi_{\alpha\dot{\gamma}} := \Psi_{\alpha\beta}\Psi_{\beta\gamma}$ are independently defined and satisfy the remaining rules of the sheaf system solution.

Remark: To continue the construction of the sheaf system resolution it may be necessary to insert more terms in the definition of the $\mathcal{R}^p_\alpha$ and the relating homomorphisms.

(2.3) Let now $\ldots \to \mathcal{R}^2_\alpha \to \mathcal{R}^1_\alpha \to \mathcal{R}^o_\alpha \to \mathcal{F}|U_\alpha \to 0$ be a resolution of $\mathcal{F}$ by sheaf systems with respect to the Stein covering $\mathfrak{U} = (U_i)_{i\in I}$. For any $\mathcal{O}_X$-module $\mathcal{G}$ we consider the dual sheaf system defined by

$$\mathcal{H}^p_\alpha = \mathcal{H}om(\mathcal{R}^p_\alpha, \mathcal{G})$$

with the transposes of the $\Phi^p_{\alpha\beta}$ as relating homomorphisms. We define

$$C^q(\mathfrak{U}, \mathcal{H}^p_*) = \prod_{\alpha\in I^{q+1}} \Gamma(U_\alpha, \mathcal{H}^p_\alpha)$$

and a coboundary operator

$$\delta : C^q(\mathfrak{U}, \mathcal{H}^p_*) \longrightarrow C^{q+1}(\mathfrak{U}, \mathcal{H}^p_*)$$

by assigning to $(h_{i_o\ldots i_q})$ the chain with $(i_o\ldots i_{q+1})$-component

$$\sum_{\nu=o}^{q+1} (-1)^\nu h_{i_o\ldots\hat{i}_\nu\ldots i_{q+1}} \circ \Phi^p_{i_o\ldots\hat{i}_\nu\ldots i_{q+1}, i_o\ldots i_q} .$$

The relation $\delta\circ\delta = 0$ follows from $\Phi^p_{\alpha\beta}\Phi^p_{\beta\gamma} = \Phi^p_{\alpha\gamma}$.
By the resolution $\mathcal{R}^\bullet_\alpha$ for each α we also have an induced differential $\partial : C^q(\mathfrak{U}, \mathcal{H}^p_*) \to C^q(\mathfrak{U}, \mathcal{H}^{p+1}_*)$, which commutes with δ. Hence we can consider the total complex $C^\bullet(\mathfrak{U}, \mathcal{F}, \mathcal{G})$, defined by

$$C^q(\mathfrak{U}, \mathcal{F}, \mathcal{G}) := \bigoplus_{i+j=q} C^i(\mathfrak{U}, \mathcal{H}om(\mathcal{R}^j_*, \mathcal{G})) ,$$

together with the total differential d as usual.

<u>(2.4) Proposition: With the above notations we have a canonical isomorphism</u> $H^q C^\bullet(\mathfrak{U}, \mathcal{F}, \mathcal{G}) \cong \mathrm{Ext}^q(X, \mathcal{F}, \mathcal{G})$.

If the resolution $\mathcal{R}^p_\alpha$ by sheaf systems is only known for $|\alpha| + p \leq 4$ then the isomorphisms still hold for $q = 0,1,2$.

We only indicate the canonical proof, the case for $|\alpha| + p \leq 4$ will follow automatically. First if $\mathcal{J}$ is any injective $\mathcal{O}_X$-module, it follows easily that $H^q C^\cdot(\mathcal{U}, \mathcal{F}, \mathcal{J}) = 0$ for $q>0$ and $H^0 C^\cdot(\mathcal{U}, \mathcal{F}, \mathcal{J}) = \mathrm{Hom}(X, \mathcal{F}, \mathcal{J})$. Now by an injective resolution $0 \longrightarrow \mathcal{G} \longrightarrow \mathcal{J}^0 \longrightarrow \mathcal{J}^1 \longrightarrow \ldots$ of $\mathcal{G}$ we obtain the double complex $c^p(\mathcal{U}, \mathcal{F}, \mathcal{J}^q)$ which turns out to be acyclic in both directions. From this we get the result by standard arguments. If the resolution terms $\mathcal{R}^p_\alpha$ are only known for $|\alpha| + p \leq 4$, then in the above complexes all modules of total order ≤ 3 are defined and the arguments work for $\mathrm{Ext}^q(X, \mathcal{F}, \mathcal{G})$ for $q = 0,1,2$.

(2.5) Corollary: If $\mathcal{F}$ has compact support and if $\mathcal{U} = (U_i)_{i\in I}$ is a finite Stein open covering of $\mathrm{Supp}(\mathcal{F})$ with a resolution $\mathcal{R}^\cdot_* \to \mathcal{F}|U_*$ with respect to $\mathcal{U}$ as above, then

$$H^q C^\cdot(\mathcal{U}, \mathcal{F}, \mathcal{G}) = \mathrm{Ext}^q(X, \mathcal{F}, \mathcal{G}).$$

For if $Y = \bigcup_{i\in I} U_i$ then by (2.4) $H^q C^\cdot(\mathcal{U}, \mathcal{F}, \mathcal{G}) = \mathrm{Ext}^q(Y, \mathcal{F}, \mathcal{G})$. But since $\mathcal{F}$ is zero outside Y it follows by the spectral sequences given $H^i(X, \mathcal{E}xt^j(\mathcal{F}, \mathcal{G}))$ and $H^i(Y, \mathcal{E}xt^j(\mathcal{F}, \mathcal{G}))$ that $\mathrm{Ext}^q(X, \mathcal{F}, \mathcal{G}) = \mathrm{Ext}^q(Y, \mathcal{F}, \mathcal{G})$.

§ 3 Obstructions for extending deformations.

In order to construct extensions of a deformation $(\mathcal{F},S)$ we investigate the obstructions to do this when a simple extension of the germ (S,s_o) is given. For the whole paragraph we consider two fixed germs (S,s_o), (T,t_o) and we assume that an epimorphism $\mathcal{O}_{T,t_o} \longrightarrow \mathcal{O}_{S,s_o}$ is given such that its kernel $\mathcal{J}$ is annihilated by the maximal ideal $\mathfrak{m}_T(t_o) \subset \mathcal{O}_{T,t_o}$. Then $\mathcal{J}$ is a finite dimensional $\mathbb{C}$-vector space with basis, say, $f_1,\ldots,f_n$, such that

$$\mathcal{J} = \bigoplus_{\nu} f_\nu \mathcal{O}_{T,t_o} \cong \mathbb{C}^n .$$

We will also assume that S and T are 0-dimensional, i.e. geometrically they are points.

If X is a complex space we get the induced sequence

$$0 \rightarrow \mathcal{J} \otimes \mathcal{O}_{X\times T} \longrightarrow \mathcal{O}_{X\times T} \longrightarrow \mathcal{O}_{X\times S} \longrightarrow 0 ,$$

and if U is a Stein open subset of X, any $g \in \Gamma(U\times T, \mathcal{J}\otimes\mathcal{O}_{X\times T})$ has a representation

$$g = \sum_{\nu=1}^{n} g_\nu f_\nu$$

with unique coefficients $g_\nu \in \Gamma(U,\mathcal{O}_X)$.

(3.1) Let us assume now that $(\mathcal{F}, S, s_o, \alpha)$ is a deformation of $\mathcal{F}_o$ with the above germ (S, s_o). To indicate the basis we write $\mathcal{F}_S = \mathcal{F}$ and use S or T as a subscript for terms which are defined on $X \times S$ or $X \times T$. It is shown in [22] that, if the extension T of S is given as above, $\mathcal{F}_S$ canonically defines classes $\xi_1,\ldots,\xi_n \in \mathrm{Ext}^2(X, \mathcal{F}_o, \mathcal{F}_o)$ and that $\mathcal{F}_S$ extends to a deformation $\mathcal{F}_T$ of $\mathcal{F}_o$ over $X \times T$ if and only if

$\xi_1 = \dots = \xi_n = 0$. In the following we are going to make explicit this statement in terms of the local resolutions of $\mathcal{F}_S$.

(3.2) With the notations above we assume that with respect to a locally finite Stein covering $\mathfrak{U}$ of X the sheaf $\mathcal{F}_S$ has a resolution by sheaf systems, cf. § 2, as follows.

$$(\#_S)\qquad \begin{array}{ccccccccc} \mathcal{R}^2_{\alpha S} & \xrightarrow[B_{\alpha S}]{} & \mathcal{R}^1_{\alpha S} & \xrightarrow[A_{\alpha S}]{} & \mathcal{R}^0_{\alpha S} & \longrightarrow & \mathcal{F}_S|U_\alpha \times S & \longrightarrow & 0 \\ \uparrow{\scriptstyle \Theta_{\alpha\beta S}} & & \uparrow{\scriptstyle \Psi_{\alpha\beta S}} & & \uparrow{\scriptstyle \Phi_{\alpha\beta S}} & & \| & & \\ \mathcal{R}^2_{\beta S} & \xrightarrow[B_{\beta S}]{} & \mathcal{R}^1_{\beta S} & \xrightarrow[A_{\beta S}]{} & \mathcal{R}^0_{\beta S} & \longrightarrow & \mathcal{F}_S|U_\beta \times S & \longrightarrow & 0 \end{array}$$

where each $\mathcal{R}^i_{\alpha S}$ is a finite direct sum of $\mathcal{O}_{U_\alpha \times S}$. We further assume that this resolution on X, i.e. modulo $\mathfrak{m}_S(s_0)$, induces a given resolution by sheaf systems of $\mathcal{F}_0$ as below:

$$(\#_0)\qquad \begin{array}{ccccccccc} \mathcal{R}^2_{\alpha} & \xrightarrow[B^0_{\alpha}]{} & \mathcal{R}^1_{\alpha} & \xrightarrow[A^0_{\alpha}]{} & \mathcal{R}^0_{\alpha} & \xrightarrow[P^0_\alpha]{} & \mathcal{F}|U_\alpha & \longrightarrow & 0 \\ \uparrow{\scriptstyle \Theta^0_{\alpha\beta}} & & \uparrow{\scriptstyle \Psi^0_{\alpha\beta}} & & \uparrow{\scriptstyle \Phi^0_{\alpha\beta}} & & \| & & \\ \mathcal{R}^2_{\beta} & \xrightarrow[B^0_{\beta}]{} & \mathcal{R}^1_{\beta} & \xrightarrow[A^0_{\beta}]{} & \mathcal{R}^0_{\beta} & \xrightarrow[P^0_\beta]{} & \mathcal{F}|U_\beta & \longrightarrow & 0\,. \end{array}$$

In the following it is sufficient to know the above resolutions only up to degree 3, cf. § 2. According to (2.3) the system $(\#_0)$ gives rise to the complex

$$C^1(\mathfrak{U}, \mathcal{F}_0, \mathcal{F}_0) \xrightarrow{d^1} C^2(\mathfrak{U}, \mathcal{F}_0, \mathcal{F}_0) \xrightarrow{d^2} C^3(\mathfrak{U}, \mathcal{F}_0, \mathcal{F}_0)$$

and explicitly we have

$$\begin{array}{ccccc}
 & & C^{o}(\mathfrak{U},\mathcal{H}om(\mathcal{R}^{1}_{*},\mathcal{F}_{o})) & \xrightarrow{\partial} & C^{o}(\mathfrak{U},\mathcal{H}om(\mathcal{R}^{2}_{*},\mathcal{F}_{o})) \\
 & & \downarrow \delta & & \\
C^{1}(\mathfrak{U},\mathcal{H}om(\mathcal{R}^{o}_{*},\mathcal{F}_{o})) & \xrightarrow{\partial} & C^{1}(\mathfrak{U},\mathcal{H}om(\mathcal{R}^{1}_{*},\mathcal{F}_{o})) & & \\
\downarrow \delta & & & & \\
C^{2}(\mathfrak{U},\mathcal{H}om(\mathcal{R}^{o}_{*},\mathcal{F}_{o})) . & & & &
\end{array} \qquad (\mathbb{K})$$

If $\xi=(\xi_{ij})\in C^{1}(\mathfrak{U},\mathcal{H}om(\mathcal{R}^{o}_{*},\mathcal{F}_{o}))$ and $\eta=(\eta_{i})\in C^{o}(\mathfrak{U},\mathcal{H}om(\mathcal{R}^{1}_{*},\mathcal{F}_{o}))$ are given, then

$$d^{1}(\xi\oplus\eta) = \mathbb{c}\oplus\mathbb{b}\oplus\mathbb{a}$$

with $\mathbb{a} = \partial\eta$, $\mathbb{b} = -\partial\xi + \delta\eta$, $\mathbb{c} = \delta\xi$ and explicitly we have the equations

(I) $\mathbb{a}_{i} = \eta_{i}B^{o}_{i}$

(II) $\mathbb{b}_{ij} = -\xi_{ij}A^{o}_{ij} + \eta_{j}\Psi^{o}_{j,ij} - \eta_{i}\Psi^{o}_{i,ij}$

(III) $\mathbb{c}_{ijk} = \xi_{jk}\Phi^{o}_{jk,ijk} - \xi_{ik}\Phi_{ik,ijk} + \xi_{ij}\Phi^{o}_{ij,ijk}$.

(3.3) Let now $M' = (A'_{\alpha T}, B'_{\alpha T}, \Phi'_{\alpha\beta T}, \Psi'_{\alpha\beta T}, \Theta'_{\alpha\beta T})$ be a system of arbitrary matrices on $(U_{\alpha}\times T)$ which induce the matrices of $(\#_{S})$ when restricted to S. Since for the matrices of $(\#_{S})$ we have all the commutativity relations, we obtain matrices R on X such that

$$\begin{array}{lll}
A'_{\alpha T}B'_{\alpha T} = \Sigma f_{\nu}R^{\nu}_{\alpha} & & \text{for } |\alpha| \leqq 3 \\
\Phi'_{\alpha\beta T}A'_{\beta T} - A'_{\alpha T}\Psi'_{\alpha\beta T} = \Sigma f_{\nu}R^{\nu}_{\alpha\beta} & & \text{for } \alpha \hookrightarrow \beta,\ |\beta| \leqq 3 \\
\Psi'_{\alpha\beta T}B'_{\beta T} - B'_{\alpha T}\Theta'_{\alpha\beta T} = \Sigma f_{\nu}R^{*\nu}_{\alpha\beta} & & \text{for } \alpha \hookrightarrow \beta,\ |\beta| \leqq 3 \\
\Phi'_{\alpha\beta T}\Phi'_{\beta\gamma T} - \Phi'_{\alpha\gamma T} = \Sigma f_{\nu}R^{\nu}_{\alpha\beta\gamma} & & \text{for } \alpha \hookrightarrow \beta \hookrightarrow \gamma,\ |\gamma|=3 \\
\Psi'_{\alpha\beta T}\Psi'_{\beta\gamma T} - \Psi'_{\alpha\gamma T} = \Sigma f_{\nu}R^{*\nu}_{\alpha\beta\gamma} & & \text{for } \alpha \hookrightarrow \beta \hookrightarrow \gamma,\ |\gamma|=3.
\end{array}$$

Now we define homomorphisms $a^{\nu}_{i}\in \mathrm{Hom}(U_{i},\mathcal{R}^{2}_{i},\mathcal{F}_{o})$, $\mathbb{b}^{\nu}_{ij}\in \mathrm{Hom}(U_{ij},\mathcal{R}^{1}_{ij},\mathcal{F}_{o})$ and $\mathbb{c}^{\nu}_{ijk}\in \mathrm{Hom}(U_{ijk},\mathcal{R}^{o}_{ijk},\mathcal{F}_{o})$ by the formulae

$$\mathbb{a}_i^\nu = -P_i^o R_i^\nu$$

$$\mathbb{b}_{ij}^\nu = P_j^o R_{j,ij}^\nu - P_i^o R_{i,ij}$$

$$\mathbb{c}_{ijk}^\nu = P_i^o(R_{i,ij,ijk}^\nu - R_{i,ik,ijk}^\nu) + P_j^o(R_{j,jk,ijk}^\nu - R_{j,ij,ijk}^\nu) + P_k^o(R_{k,ik,ijk}^\nu - R_{k,jk,ijk}^\nu) .$$

Finally setting

$$\mathbb{a}^\nu = (\mathbb{a}_i^\nu) \ , \ \mathbb{b}^\nu = (\mathbb{b}_{ij}^\nu) \ , \ \mathbb{c}^\nu = (\mathbb{c}_{ijk}^\nu)$$

we obtain the elements $\mathbb{c}^\nu \oplus \mathbb{b}^\nu \oplus \mathbb{a}^\nu \in C^2(\mathcal{U}, \mathcal{F}_o, \mathcal{F}_o)$ as indicated by the complex ($\mathbb{K}$). We claim that the following statements hold

(3.3.1) $\mathbb{c}^\nu + \mathbb{b}^\nu + \mathbb{a}^\nu \in \operatorname{Ker} d^2$.

(3.3.2) The classes $[\mathbb{c}^\nu + \mathbb{b}^\nu + \mathbb{a}^\nu] \in \operatorname{Ext}^2(X, \mathcal{F}_o, \mathcal{F}_o)$ do not depend on the choice of the system M' over the matrices of $(\#_S)$.

<u>(3.3.3) Proposition: The deformation $\mathcal{F}_S$ has an extension $\mathcal{F}_T$ as a deformation if and only if</u> $[\mathbb{c}^\nu + \mathbb{b}^\nu + \mathbb{a}^\nu] = 0$ <u>for</u> $\nu = 1,\dots,n$.

(3.4) We give a proof only for (3.3.3), the other statements can be checked in a similar and partially simpler way. Further we can assume that the ideal $\mathfrak{J} \subset \mathcal{O}_{T,t_o}$ is generated by only one element f since otherwise we only have to consider direct sums of such ideals.

(3.4.1) First let $\mathcal{F}_T$ be a deformation which extends $\mathcal{F}_S$. We have to show that $[\mathbb{c} + \mathbb{b} + \mathbb{a}] = 0$. By using Nakayama's Lemma, flatness and theorem B we find by canonical homological arguments matrices $A_{\alpha T}$, $B_{\alpha T}$, $\Phi_{\alpha\beta T}$, $\Psi_{\alpha\beta T}$ inducing $A_{\alpha S}$, $B_{\alpha S}$, $\Phi_{\alpha\beta S}$, $\Psi_{\alpha\beta S}$ when restricted to S, such that we have the commutative diagrams with exact rows:

$$\begin{array}{ccccccccc}
\mathcal{R}^2_{\alpha T} & \xrightarrow{B_{\alpha T}} & \mathcal{R}^1_{\alpha T} & \xrightarrow{A_{\alpha T}} & \mathcal{R}^0_{\alpha T} & \longrightarrow & \mathcal{F}_T|U_\alpha \times T & \longrightarrow & 0 \\
 & & \uparrow \Psi_{\alpha\beta T} & & \uparrow \Phi_{\alpha\beta T} & & \| & & \\
\mathcal{R}^2_{\beta T} & \xrightarrow{B_{\beta T}} & \mathcal{R}^1_{\beta T} & \xrightarrow{A_{\beta T}} & \mathcal{R}^0_{\beta T} & \longrightarrow & \mathcal{F}_T|U_\beta \times T & \longrightarrow & 0 .
\end{array}$$

a) For each $\alpha \hookrightarrow \beta \hookrightarrow \gamma$, $|\gamma| = 3$, there exists a holomorphic matrix $H_{\alpha\beta\gamma}$ on U_γ such that

$$R_{\alpha\beta\gamma T} := \Phi_{\alpha\beta T}\Phi_{\beta\gamma T} - \Phi_{\alpha\gamma T} = fA^0_\alpha H_{\alpha\beta\gamma} .$$

For since $\Phi_{\alpha\beta T}$ induces the identity for $\mathcal{F}_T$, we can find a matrix $M_{\alpha\beta\gamma T}$ on $U_\gamma \times T$ with $R_{\alpha\beta\gamma T} = A_{\alpha T}M_{\alpha\beta\gamma T}$, and since when restricted to S, $0 = R_{\alpha\beta\gamma S} = A_{\alpha S}M_{\alpha\beta\gamma S}$, there exists a matrix $N_{\alpha\beta\gamma S}$ on $U_\gamma \times S$ such that $M_{\alpha\beta\gamma S}=B_{\alpha S}N_{\alpha\beta\gamma S}$. If $N_{\alpha\beta\gamma T}$ is a matrix on $U_\gamma \times T$ inducing $N_{\alpha\beta\gamma S}$ on $U_\gamma \times S$, we obtain an identity $M_{\alpha\beta\gamma T} = B_{\alpha T}N_{\alpha\beta\gamma T} + fH_{\alpha\beta\gamma}$, where $H_{\alpha\beta\gamma}$ is holomorphic on U_γ. This implies $R_{\alpha\beta\gamma T} = fA_\alpha H_{\alpha\beta\gamma}$ because $f\,\mathfrak{m}_T(t_0) = 0$.

b) Since both $A'_{\alpha T}$ and $A_{\alpha T}$ induce $A_{\alpha S}$, and similarly for the other matrices, there exist matrices A_α, B_α, $\Phi_{\alpha\beta}$, $\Psi_{\alpha\beta}$ on (U_α) such that

$$A_{\alpha T} = A'_{\alpha T} + fA_\alpha \qquad \Phi_{\alpha\beta}T = \Phi'_{\alpha\beta T} + f\Phi_{\alpha\beta}$$

$$B_{\alpha T} = B'_{\alpha T} + fB_\alpha \qquad \Psi_{\alpha\beta T} = \Psi'_{\alpha\beta T} + f\Psi_{\alpha\beta} .$$

Now we define $\eta_i = P^o_i A_i$, $\xi_{ij} = P^o_j \Phi_{j,ij} - P^o_i \Phi_{i,ij}$ and claim that $d^1(\xi \oplus \eta) = \mathbb{c} \oplus \mathbb{b} \oplus \mathbb{a}$, cf. (3.). First we have

$$0 = A_{iT} B_{iT} = (A'_{iT} + fA_i)(B'_{iT} + fB_i) = f(R_i + A^o_i B_i + A_i B^o_i)$$

which implies $A_i B^o_i + A^o_i B_i = -R_i$ and hence $\eta_i B^o_i = P^o_i A_i B^o_i = -P^o_i R_i = \mathbb{a}_i$, which is (I) of (3.2) . Similarly using $\Phi_{\alpha\beta T} A_{\beta T} - A_{\beta T} \Psi_{\alpha\beta T} = 0$ we obtain

$$P^o_j R_{j,ij} - P^o_i R_{i,ij} = -(P^o_j \Phi_{j,ij} - P^o_i \Phi_{i,ij}) A^o_{ij} + (P^o_j A_j \Psi^o_{j,ij} - P^o_i A_i \Psi^o_{i,ij})$$

which is (II) of (3.2). Finally from

$$0 = \Phi_{\alpha\beta T} \Phi_{\beta\gamma T} - \Phi_{\alpha\gamma T} - fA^o_\alpha H\alpha\beta\gamma$$

we obtain

$$P^o_\alpha R_{\alpha\beta\gamma} = P^o_\alpha \Phi_{\alpha\gamma} - P^o_\alpha \Phi^o_{\alpha\beta} \Phi_{\beta\gamma} - P^o_\alpha \Phi_{\alpha\beta} \Phi^o_{\beta\gamma} \,.$$

If we insert all the terms $P^o_\alpha R_{\alpha\beta\gamma}$ into the definition of $\mathbb{c}_\gamma$, we find (III) of (3.2). This completes the proof of (3.4.1).

(3.4.2) Conversely let $\mathbb{c} + \mathbb{b} + \mathbb{a} = d^1(\xi \oplus \eta)$ with $\eta_i \in \mathrm{Hom}(U_i, \mathcal{R}^1_i, \mathcal{F}_o)$ and $\xi_{ij} \in \mathrm{Hom}(U_{ij}, \mathcal{R}^o_{ij}, \mathcal{F}_o)$. We use this equation to construct in (3.4.3) matrices A_α, B_α on U_α , $|\alpha| \leq 3$, and $\Phi_{\alpha\beta}$, $\Psi_{\alpha\beta}$ on U_β for $\alpha \hookrightarrow \beta$, $|\beta| \leq 3$, and $H_{\alpha\beta\gamma}$ on U_γ for $\alpha \hookrightarrow \beta \hookrightarrow \gamma$, $|\gamma| = 3$, such that with

$$A_{\alpha T} = A'_{\alpha T} + fA_\alpha \qquad \Phi_{\alpha\beta T} = \Phi'_{\alpha\beta T} + f\Phi_{\alpha\beta}$$

$$B_{\alpha T} = B'_{\alpha T} + fB_\alpha \qquad \Psi_{\alpha\beta T} = \Psi'_{\alpha\beta T} + f\Psi_{\alpha\beta}$$

we obtain the identities

(i) $A_{\alpha T} B_{\alpha T} = 0$ for $|\alpha| \leq 3$

(ii) $\Phi_{\alpha\beta T} A_{\beta T} = A_{\alpha T} \Psi_{\alpha\beta T}$ for $\alpha \hookrightarrow \beta$ $|\beta| \leq 3$

(iii) $\Phi_{\alpha\beta T} \Phi_{\beta\gamma T} - \Phi_{\alpha\gamma T} = A_{\alpha T} f H_{\alpha\beta\gamma}$ for $\alpha \hookrightarrow \beta \hookrightarrow \gamma$, $|\gamma| = 3$.

By (i) we obtain sheaves $\mathcal{F}_{\alpha T}$ = Coker $A_{\alpha T}$ on $U_\alpha \times T$ for $|\alpha| \leq 3$ which are T-flat, [3], [19]; note that T geometrically is a point. Hence the sheaf $\mathcal{F}_{\alpha T}$ extends the deformation $\mathcal{F}_S|U_\alpha \times S$. By (ii) homomorphisms $\varphi_{\alpha\beta}$: $\mathcal{F}_{\beta T} \to \mathcal{F}_{\alpha T}|U_\beta \times T$ for $\alpha \hookrightarrow \beta$, $|\beta| \leq 3$ are induced, which are isomorphisms. By (iii) we obtain $\varphi_{\alpha\beta}\varphi_{\beta\gamma} = \varphi_{\alpha\gamma}$. Therefore we have the following commutative diagrams for $\gamma = (i,j,k)$ of isomorphisms

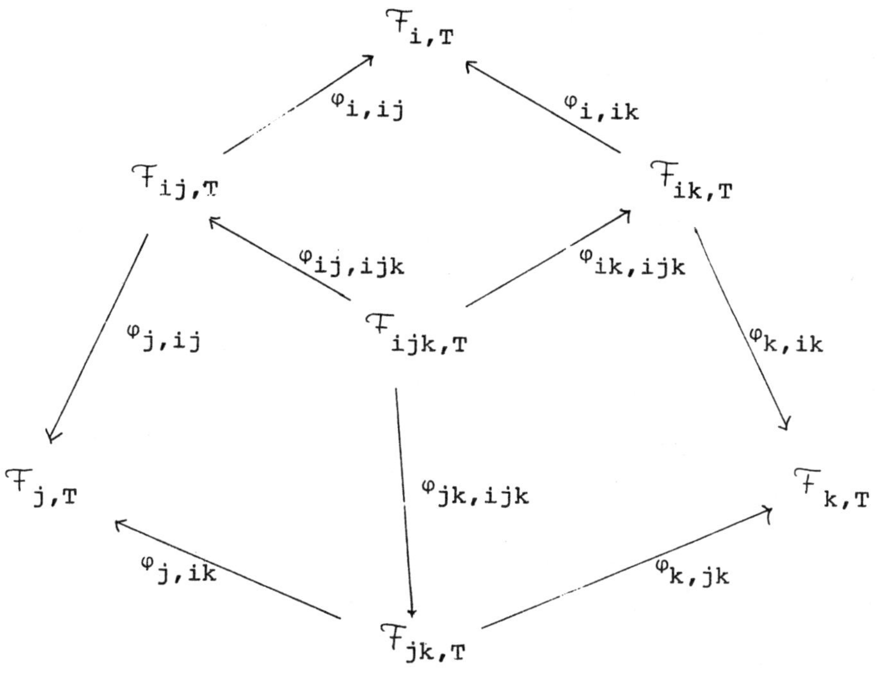

Defining σ_{ij}: $\mathcal{F}_{jT} \to \mathcal{F}_{iT}$ by $\varphi_{i,ij} \circ \varphi_{j,ij}^{-1}$ we obtain $\sigma_{ij}\,\sigma_{jk} = \sigma_{ik}$ and therefore by collation we obtain a coherent T-flat sheaf $\mathcal{F}_T$ on $X \times T$, which extends $\mathcal{F}_S$ as a deformation.

(3.4.3) Let now $\mathbb{c} \oplus \mathbb{b} \oplus \mathbb{a} = d^1(\xi \oplus \eta)$ as in (3.4.2), where $\mathbb{a}, \mathbb{b}, \mathbb{c}$ are given as in (3.3) for $n = 1$. We are going to construct the desired matrices A_α, B_α, $\Phi_{\alpha\beta}$, $\Psi_{\alpha\beta}$ and $H_{\alpha\beta\gamma}$. By the above equality the equations (I), (II), (III) of (3.2) are satisfied with some ξ_{ij} and η_i.

a) By the resolution $(\#_o)$ in (3.2) we can find a matrix A_i: $\mathcal{R}_i^1 \to \mathcal{R}_i^o$ such that $\eta_i = P_i^o A_i$. Then by (I) we get

$$- P_i^o R_i = \mathfrak{a}_i = \eta_i B_i^o = P_i^o A_i B_i^o \quad ,$$

and again by the resolution $(\#_o)$ we find a matrix B_i: $\mathcal{R}_i^2 \to \mathcal{R}_i^1$ such that

$$R_i + A_i B_i^o + A_i^o B_i = 0 \ . \qquad (1)$$

The relation (1) implies (i) for $|\alpha| = 1$.

b) As in a) we can find matrices $\Phi_{\alpha\beta}$: $\mathcal{R}_\beta^o \to \mathcal{R}_\alpha^o$ such that $\xi_{ij} = P_j^o \Phi_{j,ij} - P_i^o \Phi_{i,ij}$. (Here it is even possible to choose $\Phi_{i,ij} = 0$ or $\Phi_{j,ij} = 0$.) From (II) and the definition of $\mathfrak{b}_{ij}$ we get

$$P_j^o R_{j,ij} - P_i^o R_{i,ij} = -(P_j^o \Phi_{j,ij} - P_i^o \Phi_{i,ij}) A_{ij}^o + P_j^o A_j \Psi_{j,ij}^o - P_i^o A_i \Psi_{i,ij}^o ,$$

and from this on U_{ij}:

$$P_j^o (R_{j,ij} + \Phi_{j,ij} A_{ij}^o - A_j \Psi_{j,ij}^o) = P_i^o (R_{i,ij} + \Phi_{i,ij} A_{ij}^o - A_i \Psi_{i,ij}^o) . \qquad (2)$$

Again by (2) and the resolution $(\#_o)$ of $\mathcal{F}_o|U_{ij}$ we can find a matrix A_{ij}: $\mathcal{R}_{ij}^1 \to \mathcal{R}_{ij}^o$ such that for $\alpha \hookrightarrow \beta = (i,j)$

$$-P_\beta^o A_\beta = P_\alpha^o (R_{\alpha\beta} + \Phi_{\alpha\beta} A_\beta - A_\alpha \Psi_{\alpha\beta}^o) \quad . \qquad (3)$$

After inserting $P_\beta^o = P_\alpha^o \Phi_{\alpha\beta}^o$ we again find a matrix $\Psi_{\alpha\beta}$: $\mathcal{R}_\beta^1 \to \mathcal{R}_\alpha^1$ such that

$$R_{\alpha\beta} + \Phi_{\alpha\beta}A^{o}_{\beta} - A_{\alpha}\Psi^{o}_{\alpha\beta} + \Phi^{o}_{\alpha\beta}A_{\beta} - A^{o}_{\alpha}\Phi_{\alpha\beta} = 0 \ . \tag{4}$$

This relation implies (ii) of (3.4.2) for $\alpha \hookrightarrow \beta$, $|\beta| = 2$. Note that A_{α} for $|\alpha| \leq 2$ and $\Phi_{\alpha\beta}$, $\Psi_{\alpha\beta}$ for $\alpha \hookrightarrow \beta$ $|\beta| = 2$ have been constructed.

c) From equation (3) we obtain by multiplying from the right with B^{o}_{β} :

$$P^{o}_{\alpha}R_{\alpha\beta}B^{o}_{\beta} - P^{o}_{\alpha}A_{\alpha}\Psi^{o}_{\alpha\beta}B^{o}_{\beta} = - P^{o}_{\beta}A_{\beta}B^{o}_{\beta} \ . \tag{5}$$

Now by $f\cdot\mathfrak{m}_{T}(t_{o}) = 0$ we have the following chain of equalities for $|\beta| = 2$, cf. (3.3).

$$\begin{aligned} f\Phi^{o}_{\alpha\beta}R_{\beta} &= \Phi'_{\alpha\beta T}fR_{\beta} = \Phi'_{\alpha\beta T}A'_{\beta T}B'_{\beta T} \\ &= (A'_{\alpha T}\Psi'_{\alpha\beta T} + fR_{\alpha\beta})B'_{\beta T} \\ &= A'_{\alpha T}(B'_{\alpha T}\Theta'_{\alpha\beta T} + fR^{*}_{\alpha\beta}) + fR_{\alpha\beta}B^{o}_{\beta} \\ &= fR_{\alpha}\Phi^{o}_{\alpha\beta} + fA^{o}_{\alpha}R^{*}_{\alpha\beta} + fR_{\alpha\beta}B^{o}_{\beta} \ , \end{aligned}$$

and from this by multiplying with P^{o}_{α}:

$$P^{o}_{\beta}R_{\beta} = P^{o}_{\alpha}R_{\alpha}\Theta^{o}_{\alpha\beta} + P^{o}_{\alpha}R_{\alpha\beta}B^{o}_{\beta} \ . \tag{6}$$

From (5), (6) and (1), i.e. $P^{o}_{\alpha}R_{\alpha} = - P^{o}_{\alpha}A_{\alpha}B^{o}_{\alpha}$, we obtain

$$P^{o}_{\beta}(R_{\beta} + A_{\beta}B^{o}_{\beta}) = 0 \ .$$

and again by the resolution $(\#_{o})$ we find a matrix B_{β} such that

$$R_{\beta} + A_{\beta}B^{o}_{\beta} + A^{o}_{\beta}B_{\beta} = 0 \ . \tag{7}$$

This implies (3.4.2), (i) for $|\beta| = 2$.

d) We define the expressions

$$[i,j] := P^{o}_{j}(\Phi_{j,ij}\Phi^{o}_{ij,ijk} + R_{j,ij,ijk}) - P^{o}_{i}(\Phi_{i,ij}\Phi^{o}_{ij,ijk} + R_{i,ij,ijk})$$

for any $(i,j) = \beta \hookrightarrow \gamma = (i,j,k)$. It follows from (III) of (3.2) that we have

$$[j,k] - [i,k] + [i,j] = 0 . \tag{8}$$

Note that $[i,j] : \mathcal{R}^{o}_{ijk} \to \mathcal{F}|U_{ijk}$.

Now as in b) we can find matrices $\Phi_{i,ijk}, \Phi_{j,ijk}$ and $\Phi_{k,ijk}$ such that

$$[j,k] = P^{o}_{k}\Phi_{k,ijk} - P^{o}_{j}\Phi_{j,ijk}$$

$$[i,k] = P^{o}_{k}\Phi_{k,ijk} - P^{o}_{i}\Phi_{i,ijk} .$$

Then by (8) we also get

$$[i,j] = P^{o}_{j}\Phi_{j,ijk} - P^{o}_{i}\Phi_{i,ijk} .$$

By these equations and the definition of $[i,j]$ we can write

$$P^{o}_{j}(\Phi_{j,ij}\overset{o}{\Phi}_{ij,ijk} - \Phi_{j,ijk} + R_{j,ij,ijk}) \tag{9}$$

$$= P^{o}_{i}(\Phi_{i,ij}\overset{o}{\Phi}_{ij,ijk} - \Phi_{i,ijk} + R_{i,ij,ijk}) .$$

As before these equations show that we can find $\Phi_{\beta\gamma}$ for $\beta \hookrightarrow \gamma$ $|\beta| = 2$, $|\gamma| = 3$, such that

$$P^{o}_{\alpha}(R_{\alpha\beta\gamma} + \Phi_{\alpha\beta}\overset{o}{\Phi}_{\beta\gamma} - \Phi_{\alpha\gamma}) = - P^{o}_{\beta}\Phi_{\beta\gamma} = - P^{o}_{\alpha}\overset{o}{\Phi}_{\alpha\beta}\Phi_{\beta\gamma}$$

independently of $\alpha \hookrightarrow \beta$ with $|\alpha| = 1$. By using again the resolution $(\#_o)$ of $\mathcal{F}_o|U_\gamma$ we can find a matrix $H_{\alpha\beta\gamma}$ such that

$$R_{\alpha\beta\gamma} + \Phi_{\alpha\beta}\overset{o}{\Phi}_{\beta\gamma} + \overset{o}{\Phi}_{\alpha\beta}\Phi_{\beta\gamma} - \Phi_{\alpha\gamma} = A^{o}_{\alpha}H_{\alpha\beta\gamma} . \tag{10}$$

This relation implies (3.4.2), (iii). It remains to construct A_γ, B_γ and $\Psi_{\alpha\gamma}$, $\Psi_{\beta\gamma}$ for $|\alpha| = 1$, $|\beta| = 2$ and $|\gamma| = 3$.

e) We derive now the equation

$$R_{\alpha\gamma} + R_{\alpha\beta\gamma}A^{o}_{\gamma} = A^{o}_{\alpha}R^{*}_{\alpha\beta\gamma} + R_{\alpha\beta}\Psi^{o}_{\beta\gamma} + \Phi^{o}_{\alpha\beta} R_{\beta\gamma} \tag{11}$$

for $\alpha \hookrightarrow \beta \hookrightarrow \gamma$, $|\gamma| = 3$ similarly to c). For by $f\, \mathfrak{m}_T(t_o) = 0$ we have the identities:

$$\Phi'_{\alpha\beta T}\Phi'_{\beta\gamma T}A'_{\gamma T} = (\Phi'_{\alpha\gamma T} + fR_{\alpha\beta\gamma})A'_{\gamma T}$$

$$= A'_{\alpha T}\Psi'_{\alpha\gamma T} + fR_{\alpha\gamma} + fR_{\alpha\beta\gamma}A^o_\gamma$$

and

$$\Phi'_{\alpha\beta T}\Phi'_{\beta\gamma T}A'_{\gamma T} = \Phi'_{\alpha\beta T}(A'_{\beta T}\Psi'_{\beta\gamma T} + fR_{\beta\gamma})$$

$$= (A'_{\alpha T}\Psi'_{\alpha\beta T} + fR_{\alpha\beta})\Psi'_{\beta\gamma T} + f\Phi^o_{\alpha\beta}R_{\beta\gamma}$$

$$= A'_{\alpha T}\Psi'_{\alpha\gamma T} + fA^o_\alpha R^*_{\alpha\beta\gamma} + fR_{\alpha\beta}\Psi^o_{\beta\gamma} + f\Phi^o_{\alpha\beta}R_{\beta\gamma} .$$

f) From (4) we obtain by multiplying from the left with P^o_α and from the right with $\Psi^o_{\beta\gamma}$, where $\alpha \hookrightarrow \beta \hookrightarrow \gamma$, $|\gamma| = 3$:

$$P^o_\alpha R_{\alpha\beta}\Psi^o_{\beta\gamma} = P^o_\alpha A_\alpha \Psi^o_{\alpha\beta}\Psi^o_{\beta\gamma} - P^o_\alpha \Phi_{\alpha\beta}A^o_\beta\Psi^o_{\beta\gamma} - P^o_\alpha\Phi^o_{\alpha\beta}A_\beta\Psi^o_{\beta\gamma} \tag{12}$$

and from (10) we obtain

$$P^o_\alpha R_{\alpha\beta\gamma}A^o_\gamma = P^o_\alpha\Phi_{\alpha\gamma}A^o_\gamma - P^o_\alpha\Phi_{\alpha\beta}\Phi^o_{\beta\gamma}A^o_\gamma - P^o_\alpha\Phi^o_{\alpha\beta}\Phi_{\beta\gamma}A^o_\gamma \; . \tag{13}$$

Inserting (12), (13) into the following equation, derived from (11):

$$P^o_\alpha R_{\alpha\beta\gamma}A^o_\gamma + P^o_\alpha R_{\alpha\gamma} = P^o_\alpha R_{\alpha\beta}\Psi^o_{\beta\gamma} + P^o_\alpha\Phi^o_{\alpha\beta}R_{\beta\gamma} ,$$

we obtain for any $\alpha \hookrightarrow \beta \hookrightarrow \gamma$, $|\alpha| = 1$, $|\beta| = 2$, $|\gamma| = 3$:

$$P^o_\alpha(R_{\alpha\gamma} + \Phi_{\alpha\gamma}A^o_\gamma - A_\alpha\Psi^o_{\alpha\gamma}) = P^o_\beta(R_{\beta\gamma} + \Phi_{\beta\gamma}A^o_\gamma - A_\beta\Phi^o_{\beta\gamma}) \; . \tag{14}$$

By this equation we can find A_γ independently of α such that for any $\alpha \hookrightarrow \gamma$, $|\alpha| = 1$, $|\gamma| = 3$ we have

$$- P^o_\gamma A_\gamma = P^o_\alpha(R_{\alpha\gamma} + \Phi_{\alpha\gamma}A^o_\gamma - A_\alpha\Phi^o_{\alpha\gamma}) \; . \tag{15}$$

Since $P^o_\gamma = P^o_\alpha\Phi^o_{\alpha\gamma}$ we again can find homomorphisms $\Psi_{\alpha\gamma}$ such that

$$R_{\alpha\gamma} + \Phi_{\alpha\gamma}A^o_\gamma - A_\alpha\Psi^o_{\alpha\gamma} + \Phi^o_{\alpha\gamma}A_\gamma - A^o_\alpha\Psi_{\alpha\gamma} = 0 \; . \tag{16}$$

This equation implies (3.4.2), (ii) for $\alpha \hookrightarrow \gamma$, $|\alpha| = 1$, $|\gamma| = 3$. From (14), (15) we also have

$$- P^{O}_{\beta}\Phi^{O}_{\beta\gamma}A_{\gamma} = P^{O}_{\beta}(R_{\beta\gamma} + \Phi_{\beta\gamma}A^{O}_{\gamma} - A_{\beta}\Psi^{O}_{\beta\gamma}), \tag{17}$$

and again we can find a matrix $\psi_{\beta\gamma}: \mathcal{R}^1_{\gamma} \to \mathcal{R}^1_{\beta}$ such that

$$R_{\beta\gamma} + \Phi_{\beta\gamma}A^{O}_{\gamma} - A_{\beta}\Psi^{O}_{\beta\gamma} + \Phi^{O}_{\beta\gamma}A_{\gamma} - A^{O}_{\beta}\Psi_{\beta\gamma}, \tag{18}$$

which implies (3.4.2), (ii) for $\beta \hookrightarrow \gamma$, $|\beta| = 2$, $|\gamma| = 3$.

g) It remains to construct B_{γ}, $|\gamma| = 3$, to satisfy (3.4.2), (i) in this case. Now we have

$$\begin{aligned} f\Phi^{O}_{\alpha\gamma}R_{\gamma} &= \Phi'_{\alpha\gamma T}A'_{\gamma T}B'_{\gamma T} = (A'_{\alpha T}\Psi'_{\alpha\gamma T} + fR_{\alpha\gamma})B'_{\gamma T} \\ &= A'_{\alpha T}(B'_{\alpha T}\Theta'_{\alpha\gamma T} + fR^{*}_{\alpha\gamma}) + fR_{\alpha\gamma}B^{O}_{\gamma} \\ &= fR_{\alpha}\Theta^{O}_{\alpha\gamma} + fA^{O}_{\alpha}R^{*}_{\alpha\gamma} + fR_{\alpha\gamma}B^{O}_{\gamma}, \end{aligned}$$

by which we get

$$P^{O}_{\gamma}R_{\gamma} = P^{O}_{\alpha}R_{\alpha}\Theta^{O}_{\alpha\gamma} + P^{O}_{\alpha}R_{\alpha\gamma}B^{O}_{\gamma}. \tag{19}$$

However by the definition of A_{γ} in (15) we have

$$\begin{aligned} - P^{O}_{\gamma}A_{\gamma}B^{O}_{\gamma} &= P^{O}_{\alpha}R_{\alpha\gamma}B^{O}_{\gamma} - P^{O}_{\alpha}A_{\alpha}\Psi^{O}_{\alpha\gamma}B_{\gamma} \\ &= P^{O}_{\alpha}(R_{\alpha\gamma}B^{O}_{\gamma} - A_{\alpha}B^{O}_{\alpha}\Theta^{O}_{\alpha\gamma}) \qquad \text{and by (1)} \\ &= P^{O}_{\alpha}(R_{\alpha\gamma}B^{O}_{\gamma} + R_{\alpha}\Theta^{O}_{\alpha\gamma}) \\ &= P^{O}_{\gamma}R_{\gamma} : \end{aligned}$$

Hence there exists B_{γ} with

$$R_{\gamma} + A_{\gamma}B^{O}_{\gamma} + A^{O}_{\gamma}B_{\gamma} = 0,$$

and this finally implies (3.4.2), (i) for $|\gamma| = 3$. This completes the proof of (3.4.2) and (3.3.3).

(3.5) Remark: If in the diagram ($\#_S$) of (3.2) for any $\beta = (i,j)$ we have $\Phi_{i,ij,S}(x \oplus y) = x$, we can assume that also $\Phi_{i,ij,T}(x \oplus y) = x$ in the construction (3.4.2), (3.4.3). For since $\Phi_{i,ij,S}$ is constant, we can assume that also $\Phi'_{i,ij,T}(x \oplus y) = x$, and by choosing $\phi_{i,ij} = 0$ in (3.4.3), b) we get $\Phi_{i,ij,T} = \Phi'_{i,ij,T}$. This will be used in (10.12) and (10.13).

§ 4 Amalgamation

Here some auxiliary Lemmas on the amalgamation of local resolutions without bounds are proved which will be used later to shorten the constructions with bounds.

(4.1) Amalgamation I: Let X be a Stein space and S a one-point space with possibly unreduced structure. Let $\mathcal{U} = (U_i)$ be a Stein covering of X and assume that matrices σ_1, σ_2 on $X\times S$ and φ_i on $U_i\times S$ are given in the following diagram

$$\begin{array}{ccccccc} \mathcal{O}^{q_1}_{X\times S} & \xrightarrow[\sigma_1]{} & \mathcal{O}^{p_1}_{X\times S} & \longrightarrow & \operatorname{Coker}\sigma_1 & \longrightarrow & 0 \\ & & \Big\downarrow \varphi_i & & & & \\ \mathcal{O}^{q_2}_{X\times S} & \xrightarrow[\sigma_2]{} & \mathcal{O}^{p_2}_{X\times S} & \longrightarrow & \operatorname{Coker}\sigma_2 & \longrightarrow & 0\ . \end{array}$$

We assume further that the φ_i induce isomorphisms $\overline{\varphi}_i$ for the cokernels and that $\overline{\varphi}_i = \overline{\varphi}_j$ on $U_{ij}\times S$.
Then there are homomorphisms

$$\mathcal{O}^{p_1}_{X\times S} \xrightarrow[\varphi]{} \mathcal{O}^{p_2}_{X\times S} \quad \text{and} \quad \mathcal{O}^{p_2}_{X\times S} \xrightarrow[\psi]{} \mathcal{O}^{p_1}_{X\times S}$$

inducing isomorphisms $\overline{\varphi}$ and $\overline{\psi}$ for the cokernels such that

$$\overline{\psi} = \overline{\varphi}^{-1} \quad \text{and} \quad \overline{\varphi}|U_i\times S = \overline{\varphi}_i\ .$$

Proof: a) Let e_ν denote the νth basis element of $\Gamma(X\times S, \mathcal{O}^{p_1}_{X\times S})$ and let $\xi_{ij\nu} \in \Gamma(U_{ij}\times S,\ \operatorname{Im}\sigma_2)$ be defined by

$$\xi_{ij\nu} = \varphi_j(e_\nu) - \varphi_i(e_\nu)\ .$$

Then by $\xi_\nu = (\xi_{ij\nu})_{i,j}$ we have a cocycle in $Z^1(\mathcal{U}\times S,\ \operatorname{Im}\sigma_2)$

and since X is Stein there exist $\eta_{i\nu} \in \Gamma(U_i \times S, \operatorname{Im} \sigma_2)$ such that

$$\xi_{ij\nu} = \eta_{j\nu} - \eta_{i\nu} .$$

Hence by $\varphi(e_\nu) = \varphi_i(e_\nu) - \eta_{i\nu}$ a global homomorphism φ is defined. By its definition φ induces a homomorphism $\overline{\varphi}$ for the cokernels and moreover $\overline{\varphi} = \overline{\varphi}_i$ on $U_i \times S$ because $\overline{\eta}_i = 0$.

b) To prove the existence for ψ we consider the induced homomorphism (φ, σ_2)

$$\mathcal{O}^{p_1}_{X\times S} \oplus \mathcal{O}^{q_2}_{X\times S} \underset{\binom{\psi}{g}}{\overset{(\varphi,\sigma_2)}{\rightleftarrows}} \mathcal{O}^{p_2}_{X\times S}$$

which is surjective since $\overline{\varphi}$ is an isomorphism. By choosing elements which are mapped to the canonical basis sections of $\mathcal{O}^{p_2}_{X\times S}$ we obtain a map $\binom{\psi}{g}$ such that

$$\varphi\ \psi + \sigma_2 g = (\varphi, \sigma_2)\binom{\psi}{g} = \mathrm{id} .$$

Now ψ induces a homomorphism $\overline{\psi}$ for the cokernels: For if $s \in \operatorname{Im} \sigma_2$ then $\varphi \circ \psi(s) = s - \sigma_2 g(s)$ and hence $\overline{\varphi}\ \overline{\psi(s)} = 0$. Since $\overline{\varphi}$ is bijective, $\overline{\psi(s)} = 0$, i.e. $\psi(s) \in \operatorname{Im} \sigma_1$. Now by definition $\overline{\varphi} \circ \overline{\psi} = \mathrm{id}$.

<u>(4.2) Definition:</u> We identify $\mathbb{C}^N$ and $\mathbb{R}^{2N}$ as usual and consider open rectangles $Q = I_1 \times \ldots \times I_{2N}$ where I_ν are open intervals of the real line. Two such open rectangles $Q_i = I^i_1 \times \ldots \times I^i_{2N}$ are called amalgamatable if there exists ν such that: $I^1_\nu \cap I^2_\nu \neq \emptyset$ and $I^1_\mu = I^2_\mu$ for $\mu \neq \nu$.

(4.3) Amalgamation II: Let Q_1, Q_2 be open amalgamatable rectangles in $\mathbb{C}^N$ and let $X \subset Q_1 \cup Q_2$ be a closed analytic subspace, and let $Q_{12} = Q_1 \cap Q_2$, $X_i = X \cap Q_i$, $X_{12} = X_1 \cap X_2$. Let further S be a one-point space with possibly unreduced structure. Let homomorphisms be given as in the following diagram

$$\begin{array}{ccccccc}
\mathcal{O}^{q_1}_{X\times S} & \xrightarrow[\sigma_1]{} & \mathcal{O}^{p_1}_{X\times S} & \longrightarrow & \operatorname{Coker}\sigma_1 & \longrightarrow & 0 \\
 & & \varphi_{21}\big\downarrow\big\uparrow\varphi_{12} & & \overline{\varphi}_{21}\big\downarrow\big\uparrow\overline{\varphi}_{12} & & \\
\mathcal{O}^{q_2}_{X\times S} & \xrightarrow[\sigma_2]{} & \mathcal{O}^{p_2}_{X\times S} & \longrightarrow & \operatorname{Coker}\sigma_2 & \longrightarrow & 0 ,
\end{array}$$

such that σ_i is defined on $X_i\times S$ and φ_{ij} on $X_{12}\times S$, and assume that φ_{ij} induce isomorphisms $\overline{\varphi}_{ij}$ which are inverse to each other.

Claim: There are homomorphisms σ, τ_i, φ_i as in the diagram below, which are obtained by the following algebraic operations and where σ is defined over $X\times S$ and τ_i, φ_i over $X_i\times S$. Moreover τ_i, φ_i induce isomorphisms $\overline{\tau}_i$, $\overline{\varphi}_i$ for the cokernels such that $\overline{\tau}_i = \overline{\varphi}_i^{-1}$ and $\overline{\varphi}_j\overline{\varphi}_{ji} = \overline{\varphi}_i$.

$$\begin{array}{ccccccc}
\mathcal{O}^{q}_{X\times S} & \xrightarrow{\sigma} & \mathcal{O}^{p}_{X\times S} & \longrightarrow & \operatorname{Coker}\sigma & \longrightarrow & 0 \\
 & & \tau_i\big\uparrow\big\downarrow\varphi_i & & \overline{\tau}_i\big\uparrow\big\downarrow\overline{\varphi}_i & & \\
\mathcal{O}^{q_i}_{X\times S} & \xrightarrow[\sigma_i]{} & \mathcal{O}^{p_i}_{X\times S} & \longrightarrow & \operatorname{Coker}\sigma_i & \longrightarrow & 0 \quad .
\end{array}$$

Proof: Construct first ψ_{ji} and f_{ji} on $X_{12}\times S$ as follows:

$$\begin{array}{ccc}
\cdot & \xrightarrow{\sigma_i} & \cdot \\
\psi_{ji}\big\downarrow & & \big\downarrow\varphi_{ji} \\
\cdot & \xrightarrow[\sigma_j]{} & \cdot
\end{array}
\qquad\qquad
\begin{array}{ccc}
 & & \cdot \\
 & \swarrow f_{ij} & \big\downarrow \mathrm{id}-\varphi_{ij}\varphi_{ji} \\
\cdot & \xrightarrow[\sigma_i]{} & \cdot
\end{array}$$

This is possible because $\bar{\varphi}_{ji}$ exists and has a inverse $\bar{\varphi}_{ij}$. Define now

$$\hat{\varphi}_{ji} := \begin{pmatrix} \varphi_{ji} & id-\varphi_{ji}\varphi_{ij} \\ id & -\varphi_{ij} \end{pmatrix}$$

$$\hat{\psi}_{ji} := \left(\begin{array}{cc|cc} \psi_{ji} & f_{ij} & id-\psi_{ji}\psi_{ij}-f_{ij}\sigma_j & f_{ij}\varphi_{ji}-\psi_{ji}f_{ji} \\ \sigma_i & -\varphi_{ij} & o & o \\ \hline id & o & -\psi_{ij} & -f_{ji} \\ o & id & -\sigma_j & \varphi_{ji} \end{array}\right)$$

Then the following diagram is commutative.

$$\begin{array}{ccc} (\mathcal{O}^{q_i}_{X\times S} \oplus \mathcal{O}^{p_j}_{X\times S}) \oplus (\mathcal{O}^{q_j}_{X\times S} \oplus \mathcal{O}^{p_i}_{X\times S}) & \xrightarrow{(\sigma_i \oplus id, o)} & \mathcal{O}^{p_i}_{X\times S} \oplus \mathcal{O}^{p_j}_{X\times S} \\ \downarrow \hat{\psi}_{ji} & & \downarrow \hat{\varphi}_{ji} \\ (\mathcal{O}^{q_j}_{X\times S} \oplus \mathcal{O}^{p_i}_{X\times S}) \oplus (\mathcal{O}^{q_i}_{X\times S} \oplus \mathcal{O}^{p_j}_{X\times S}) & \xrightarrow[(\sigma_j \oplus id, o)]{} & \mathcal{O}^{p_j}_{X\times S} \oplus \mathcal{O}^{p_i}_{X\times S} \end{array} .$$

By direct computation we have $\hat{\psi}_{ji}\hat{\psi}_{ij} = id$ and $\hat{\varphi}_{ji}\hat{\varphi}_{ij} = id$ such that these matrices are invertible on $X_{12}\times S$. We set

$$q = q_i + q_j + p_i + p_j \ , \quad p = p_i + p_j$$

and by Cartans-Lemma, (6.11), we find invertible matrices Φ_i and Ψ_i on $X_i\times S$ such that

$$\hat{\varphi}_{ji} = \Phi_j^{-1}\Phi_i \quad \text{and} \quad \hat{\psi}_{ji} = \Psi_j^{-1}\Psi_i \ .$$

Finally we define

$$\sigma := \Phi_i(\sigma_i \oplus id, o)\Psi_i^{-1}$$

on $X_i \times S$ such that σ is defined on $X \times S$ and we define

$$\varphi_i : \mathcal{O}_{X\times S}^{p_i} \xrightarrow[\varepsilon]{} \mathcal{O}_{X\times S}^{p_i} \oplus \mathcal{O}_{X\times S}^{p_j} \xrightarrow[\Phi_i]{} \mathcal{O}_{X\times S}^{p}$$

$$\tau_i : \mathcal{O}_{X\times S}^{p} \xrightarrow[\Phi_i^{-1}]{} \mathcal{O}_{X\times S}^{p_i} \oplus \mathcal{O}_{X\times S}^{p_j} \xrightarrow[pr]{} \mathcal{O}_{X\times S}^{p_i} .$$

It is now straightforward to verify the rest of the statement.

§ 5 Reduction tools

Together with the power series method to construct analytic deformations we need the tools of Grauert's reduction theory and use these tools in the form they have been given by Forster-Knorr in [6]. For the convenience of the reader we collect the necessary formulas here and refer to [6], [11], [9] for proofs. At the end of this paragraph some auxiliary Lemmas will be proved.

(5.1) Let $\mathbb{R}_+^m$ be the set of all m-tuples $\rho = (\rho_1, \ldots, \rho_m)$ of real numbers $\rho_i > 0$. Let $\gamma(t) = \sum_{|\nu| \geq 0} \gamma_\nu t^\nu$ be a convergent power series with real coefficients such that

(a) $\sum_{\nu+\mu=\lambda} \gamma_\nu \gamma_\mu \leq \gamma_\lambda$

(b) $\gamma_\nu \geq \gamma_{\nu+\mu}$

(c) $\rho^\nu \leq c\gamma_\nu$ for some $c > 0$, $\rho \in \mathbb{R}_+^m$.

For example the coefficients of $\gamma(t) = \prod_{i=1}^{m} (\frac{1}{2} - \frac{1}{4}\sqrt{1-t_i})$ will satisfy these conditions. We will keep fixed such a "majorant" γ for the whole paper. By condition (c) we have $\gamma_\nu > 0$ for all ν.

If E is any $\mathbb{C}$-Banach space let

$$E\langle t\rangle = E\langle t_1, \ldots, t_m\rangle$$

be the space of convergent power series with coefficients in E. If $f = \sum_{|\nu| \geq 0} f_\nu t^\nu$ is such a series, we define

$$\|f\|_\rho = \sup_{|\nu| \geq 0} \|f_\nu\| \frac{\rho^\nu}{\gamma_\nu} .$$

This norm needs not be finite but the elements of $E\langle t\rangle$ with finite norm form a Banach space. Moreover we have

(i) $\|t^{\nu}f\|_{\rho} \geqq \rho^{\nu}\|f\|_{\rho}$

(ii) If $E \times F \to G$ is a bilinear mapping $(x,y) \to xy$ with $\|xy\| \leqq \|x\| \, \|y\|$ and if $E<t> \times F<t> \to G<t>$ is the canonically induced bilinear mapping, then for $f \in E<t>$, $g \in F<t>$ we have $\|fg\|_{\rho} \leqq \|f\|_{\rho}\|g\|_{\rho}$ by using (a).

(5.2) In the set $\mathbb{N}^m$, where $\mathbb{N}$ denotes the set of all non-negative integers, we introduce the lexicographical order as usual:

If $\nu = (\nu_1,\ldots,\nu_m)$ and $\mu = (\mu_1,\ldots,\mu_m)$ then $\nu<\mu$ iff $|\nu| < |\mu|$ or if $|\nu| = |\mu|$ and for some $k \leqslant m$ we have $\nu_k<\mu_k$ but $\nu_i = \mu_i$ for $i>k$.

If E is a Banach space and $f = \Sigma f_{\nu}t^{\nu} \in E<t>$ we define ord $f = \infty$ if $f = 0$ and

$$\text{ord } f = \min \{\nu \in \mathbb{N}^m \,|\, f_{\nu} \neq 0\}$$

otherwise. As in [11] one obtains

(5.2.1) Let $(f_i)_{i \in I}$ be a finite family of elements of $E<t>$ with ord $f_i > \nu_i$, $\nu_i \in \mathbb{N}^m$. Then for any $\varepsilon > 0$ there exists $\rho \in \mathbb{R}_+^m$ such that $\|f_i\|_{\rho} \leqq \varepsilon\rho^{\nu_i}$ for $i \in I$.

(5.3) We set $\mathbb{C}_{\infty}<t> = \mathbb{C}<t> = \mathbb{C}<t_1,\ldots,t_m>$ and for $e \in \mathbb{N}$ we set $\mathbb{C}_e<t> = \mathbb{C}<t>/\mathfrak{m}^{e+1}$ where $\mathfrak{m}$ denotes the maximal ideal of $\mathbb{C}<t>$. An element $f \in \mathbb{C}_e<t>$ is also considered as a polynomial of degree $\leq$ e. If $\mathfrak{A} \subset \mathbb{C}_e\langle t\rangle$, $e \in \mathbb{N} \cup \{\infty\}$, is any ideal we define $\operatorname{ord} \mathfrak{A} = \{\operatorname{ord} f \mid f \in \mathfrak{A}\}$. The reducing system $\Lambda(\mathfrak{A})$ of $\mathfrak{A}$ is defined as the (finite) set of all $\lambda \in \operatorname{ord} \mathfrak{A}$ such that $\lambda \notin \mu + \mathbb{N}^m$ for any other $\mu \in \operatorname{ord} \mathfrak{A}$. We set

$$\Delta(\mathfrak{A}) = \mathbb{N}^m \setminus \bigcup_{\lambda \in \Lambda(\mathfrak{A})} (\lambda + \mathbb{N}^m) .$$

If E is a Banach space, a series $f = \Sigma f_\nu t^\nu$ in $E\langle t\rangle$ is called $\Lambda(\mathfrak{A})$-reduced ($\Lambda(\mathfrak{A})$-e-reduced) if $f_\nu = o$ for $\nu \notin \Delta(\mathfrak{A})$ ($\nu \notin \Delta(\mathfrak{A})$ or $e < |\nu|$). By $\widetilde{\mathfrak{A}} \subset \mathbb{C}\langle t\rangle$ we denote the inverse image of $\mathfrak{A}$ under the canonical projection $\mathbb{C}\langle t\rangle \to \mathbb{C}_e\langle t\rangle$, such that $\mathfrak{m}^{e+1} \subset \widetilde{\mathfrak{A}}$. We have, [6] :

(5.4) If $f \in \widetilde{\mathfrak{A}} E\langle t\rangle$ and if f is $\Lambda(\mathfrak{A})$-e-reduced then $f = o$.

(5.5) Grauert's reduction and division theorem:

Let $e \in \mathbb{N} \cup \{\infty\}$ and let $\mathfrak{A} \subset \mathbb{C}_e\langle t\rangle$ be any ideal. Then there is a unique system $(\omega_\lambda)_{\lambda \in \Lambda(\mathfrak{A})}$ of power series $\omega_\lambda = t^\lambda + a_\lambda \in \mathfrak{A}$ such that a_λ is $\Lambda(\mathfrak{A})$-e-reduced, $\operatorname{ord} a_\lambda > \lambda$, and such that the family (ω_λ) generates $\mathfrak{A}$. Moreover for any $f \in E\langle t\rangle$ we have a unique division

$$f = \sum_{\lambda \in \Lambda(\mathfrak{A})} q_\lambda \omega_\lambda + r$$

with q_λ, $r \in E\langle t\rangle$, $\operatorname{ord} q_\lambda + \lambda \geq \operatorname{ord} f$, $\operatorname{ord} r \geq \operatorname{ord} f$ and r $\Lambda(\mathfrak{A})$-reduced. The family $(\omega_\lambda)_{\lambda \in \Lambda(\mathfrak{A})}$ is called the Grauert-Weierstraß-family (in short, the GW-family) of .

(5.5.1) Subtracting from r all terms of total order > e (if e is finite) we obtain a unique element $\operatorname{red}_{\mathfrak{A}} f \in E\langle t\rangle$ such that

(i) $\text{red}_{\mathfrak{A}} f$ is $\Lambda(\mathfrak{A})$-e-reduced

(ii) $f \equiv \text{red}_{\mathfrak{A}} f \bmod \tilde{\mathfrak{A}}\, E\langle t\rangle$

(iii) $\text{ord red}_{\mathfrak{A}} f \geq \text{ord } f$

(iv) If $\|a_\lambda\|_\rho \leq \frac{1}{2 \text{ card } \Lambda(\mathfrak{A})} \rho^\lambda$ for every $\lambda \in \Lambda(\mathfrak{A})$ then $\|\text{red}_{\mathfrak{A}} f\|_\rho \leq 2\|f\|_\rho$.

<u>(5.6) Further notations and extensions of ideals:</u> Let $e_o < e \leq \infty$ be given and let $\mathfrak{J}_d \subset \mathbb{C}_d\langle t\rangle$ for $d \leq e$ be a family of ideals with $\mathfrak{J}_d|\mathbb{C}_{d-1}\langle t\rangle = \mathfrak{J}_{d-1}$, where $\mathfrak{J}_d|\mathbb{C}_{d-1}\langle t\rangle$ denotes the image of $\mathfrak{J}_d$ under the canonical projection $\mathbb{C}_d\langle t\rangle \to \mathbb{C}_{d-1}\langle t\rangle$. Assume further that for $e_o \leq d \leq e$ we have $\Lambda(\mathfrak{J}_d) = \Lambda(\mathfrak{J}_{e_o}) = \Lambda$. Let Δ be the set of Λ-reduced indices, cf. (5.3), and set

$$\Delta_d = \{\nu \in \Delta \mid |\nu| = d\}.$$

If ω_λ^d, $\lambda \in \Lambda$, denotes the GW-familiy of $\mathfrak{J}_d$ we have by uniqueness that

$$\omega_\lambda^d|\mathbb{C}_{d'}\langle t\rangle = \omega_\lambda^{d'}$$

for $e_o \leq d' < d \leq e$. Let now $\Lambda_o \subset \Lambda$ be minimal such that the elements $\omega_\lambda^{e_o}$, $\lambda \in \Lambda_o$, generate $\mathfrak{J}_{e_o}$. Then there are polynomials $c_{\mu\lambda}$ with $c_{\mu\lambda} = \delta_{\mu\lambda}$ for $\mu \in \Lambda_o$ such that

$$\omega_\mu^{e_o} \equiv \sum_{\lambda\in\Lambda_o} c_{\mu\lambda}\omega_\lambda^{e_o} \bmod \mathfrak{m}^{e_o+1} .$$

We define for any $b = (b_{\lambda\nu})_{\lambda \in \Lambda_o,\ \nu \in \Delta_{e+1}}$

$$\omega_\lambda^{e+1}(b) = \omega_\lambda^e + \sum_\nu b_{\lambda\nu} t^\nu \qquad \text{for } \lambda \in \Lambda_o \ ,$$

and we denote by $\mathfrak{J}_{e+1}(b)$ the ideal in $\mathbb{C}_{e+1}\langle t\rangle$ generated by those elements. The following statements with the above notations are proved in [6] :

(5.6.1) <u>If there exists an ideal</u> $\mathcal{J}_* \subset \mathbb{C}_{e+1}\langle t\rangle$ <u>with</u> $\mathcal{J}_*|\mathbb{C}_e\langle t\rangle = \mathcal{J}_e$ <u>and</u> $\Lambda(\mathcal{J}_*) = \Lambda$, <u>then</u> $\Lambda(\mathcal{J}_{e+1}(b)) = \Lambda$. <u>Moreover, if</u> $\mathcal{J} \subset C_{e+1}\langle t\rangle$ <u>is any ideal with</u> $\mathcal{J}|\mathbb{C}_e\langle t\rangle = \mathcal{J}_e$ <u>and</u> $\Lambda(\mathcal{J}) = \Lambda$, then $\mathcal{J} = \mathcal{J}_{e+1}(b)$ <u>for some</u> $b = (b_{\lambda\nu})$.

(5.6.2) Let $\omega_\mu^{e+1} = t^\mu + a_\mu$, $\mu \in \Lambda$, <u>be the GW-family of</u> $\mathcal{J}_{e+1}(b)$ <u>and assume that the following estimates are satisfied:</u>

$$\|\omega_\lambda^e - \omega_\lambda^{e_o}\|_\rho\ ,\ \|\sum_\nu b_{\lambda\nu}t^\nu\|_\rho \leq \frac{1}{8C_*\mathrm{card}(\Lambda)}|\rho|^{e_o} \text{ for } \lambda \in \Lambda_o$$

$$\|\sum_{\lambda\in\Lambda_o} c_{\mu\lambda}\omega_\lambda^{e_o} - t^\mu\|_\rho \leq \frac{1}{8\ \mathrm{card}(\Lambda)}\rho^\mu \text{ for } \mu \in \Lambda,$$

<u>where</u> $|\rho| = \min \rho_i$ <u>and</u> $C_* = \max_{\mu\in\Lambda} \sum_{\lambda\in\Lambda_o} \|c_{\mu\lambda}\|_{(1,\ldots,1)}$.

<u>Then we have</u>

(i) $\omega_\lambda^{e+1} = \omega_\lambda^{e+1}(b)$ for $\lambda \in \Lambda_o$

(ii) $\|\omega_\lambda^{e+1} - \omega_\lambda^{e_o}\|_\rho \leq \frac{1}{8C_*\ \mathrm{card}(\Lambda)}|\rho|^{e_o}$ for $\lambda \in \Lambda_o$

(iii) $\|a_\mu\|_\rho \leq \frac{1}{2\ \mathrm{card}(\Lambda)}\rho^\mu$ for $\mu \in \Lambda$.

(5.6.3) <u>Assume that the condition of (5.6.1) is satisfied and let</u> $f \in E\langle t\rangle$, E <u>a Banach space, have a decomposition</u>

$$f = \sum_{\mu\in\Lambda} Q_\mu \omega_\mu^{e_o} + r \qquad \mathrm{mod}\ \mathfrak{m}^{e_o+1}$$

<u>by the Grauert division theorem. If we write</u>

$$\mathrm{red}_{\mathcal{J}_{e+1}(b)} f = \sum_{\substack{\nu\in\Delta \\ |\nu|\leq e+1}} R_\nu^b t^\nu$$

<u>then</u> $R_\nu^o = R_\nu^b$ for $|\nu| \leq e$, <u>and for</u> $\nu \in \Delta_{e+1}$ <u>we have</u>

$$F_\nu^o - R_\nu^b = \sum_{\substack{\mu\in\Lambda \\ \lambda\in\Lambda_o}} Q_\mu(o)\, c_{\mu\lambda}(o)\, b_{\lambda\nu}.$$

(Here $Q_\mu(o)$ and $c_{\mu\lambda}(o)$ are the values of the polynomials at $t = o$).

<u>(5.6.4) Assume further that the condition of (5.6.1) is satisfied such that</u> $\Lambda(\mathfrak{J}_{e+1}(b)) = \Lambda$. <u>By definition we have</u> $\mathfrak{m}^{e+1} \subset \tilde{\mathfrak{J}}_e$ <u>and</u> $\mathfrak{m}\tilde{\mathfrak{J}}_e \subset \mathfrak{J}_{e+1}(b) \subset \tilde{\mathfrak{J}}_e$. <u>It follows from (5.4) that the monomials</u> $t^\nu, \nu\in\Delta_{e+1}$, <u>form a basis of the $\mathbb{C}$-vectorspace</u> $\tilde{\mathfrak{J}}_e/\widetilde{\mathfrak{J}_{e+1}(b)}$.

For later use we derive a number of estimates which will often be used.

(5.7) Let $\mathfrak{J}_e$ be a sequence of ideals $\mathfrak{J}_e \subset \mathbb{C}_e\langle t\rangle$ such that $\mathfrak{J}_e|\mathbb{C}_{e-1}\langle t\rangle = \mathfrak{J}_{e-1}$, $e = 1,2,\ldots$, and assume that $\Lambda(\mathfrak{J}_e) = \Lambda(\mathfrak{J}_{e_o}) = \Lambda$ for $e\geq e_o$. Let $\rho_o\in\mathbb{R}_+^m$ and assume ρ_o is choosen in such a way that

$$\| \mathrm{red}_{\mathfrak{J}_e}\, a\|_{\rho_o} \leq 2\|a\|_{\rho_o}$$

for any $e\geq e_o$ and any $a\in E\langle t\rangle$, E a Banach-space, cf.(5.5.1),(iv). We use the notations $\rho=\theta\rho_o$ and $\mathrm{red}_e=\mathrm{red}_{\mathfrak{J}_e}$. Then the following holds:

<u>Lemma:</u> <u>If</u> $a\in E\langle t\rangle$ <u>with</u> $\|a\|_{\rho_o} < \infty$ <u>and</u> $a \equiv o \bmod \mathfrak{J}_{e_o}$, <u>then for</u> $\varepsilon>0$ <u>there exists</u> $\theta_\varepsilon>o$ <u>such that for</u> $o<\theta\leq\theta_\varepsilon$

$$\| \mathrm{red}_e a\|_\rho \leq \varepsilon\, |\rho|^{e_o}$$

<u>independent of</u> $e\leq e_o$.

Proof: Let $e \gg e_o$ and write $\mathrm{red}_e\, a = \sum_{\substack{\nu\in\Delta \\ |\nu|\leq e}} a_\nu t^\nu$.

By definition $a \equiv \mathrm{red}_{e_o}\, a \mod \tilde{\mathfrak{I}}_{e_o}$ and $a \equiv \mathrm{red}_e\, a \mod \tilde{\mathfrak{I}}_e$.
Since $\tilde{\mathfrak{I}}_e \subset \tilde{\mathfrak{I}}_{e_o}$ we obtain

$$0 \equiv \mathrm{red}_{e_o}\, a \equiv \mathrm{red}_e\, a \mod \tilde{\mathfrak{I}}_{e_o}$$

and hence $a_\nu = o$ for $|\nu| \leq e_o$.
Let now θ be so small that $2\theta\|a\|_{\rho_o} \leq \varepsilon|\rho_o|^{e_o}$. We can assume that $\rho_1^o = |\rho_o| = \min \rho_i^o$. Then with $\rho = \theta\rho_o$ we obtain

$$\begin{aligned}\|\mathrm{red}_e\, a\|_\rho &= \sup\{|a_\nu|\theta^{|\nu|}\frac{\rho_o^\nu}{\gamma_\nu},\ e_o < |\nu| \leq e,\ \nu \in \Delta\} \\ &\leq \theta^{e_o+1}\|\mathrm{red}_e\, a\|_{\rho_o} \\ &\leq 2\theta^{e_o+1}\|a\|_{\rho_o} \leq \varepsilon\,\theta^{e_o}|\rho_o|^{e_o} = \varepsilon|\rho|^{e_o}.\end{aligned}$$

<u>(5.7.1) Corollary:</u> <u>If</u> $a \in E\langle t\rangle$ and $\|a\|_{\rho_o} < \infty$, <u>then for any</u> $\varepsilon > o$ <u>there exists</u> θ_ε, <u>such that for</u> $o < \theta \leq \theta_\varepsilon$

$$\|\mathrm{red}_e\, a - \mathrm{red}_{e_o}\, a\|_\rho \leq \varepsilon|\rho|^{e_o}$$

<u>independent of</u> $e \geq e_o$.

Proof: Let $b = \mathrm{red}_e\, a - \mathrm{red}_{e_o}\, a$. Then $b \equiv o \mod \tilde{\mathfrak{I}}_{e_o}$ and $\mathrm{red}_e\, b = b$ and we can apply (5.7).

(5.8) Let the sequence of ideals $\mathfrak{I}_e$ and ρ_o be as in (5.7). Let $E\times F \to G$ be a bilinear map of Banach spaces and let $L_1, L_2 > o$ be constants and $\varepsilon > o$ with $2\varepsilon(L_1 + L_2 + L_1 L_2) \leq 1$. Let $a \in E\langle t\rangle$ and $b \in F\langle t\rangle$ be Λ-reduced series with finite ρ_o-norm and let a_e, b_e be their truncations mod $\mathfrak{m}^{e+1}$. By (5.7.1) we can choose θ_ε such that for $o < \theta \leq \theta_\varepsilon$ and $e \geq e_o$

$$\| \text{red}_e a_{e_o} b_{e_o} - \text{red}_{e_o} a_{e_o} b_{e_o} \|_\rho \leq \varepsilon \, |\rho|^{e_o}$$

and in addition

$$\| a_{e_o} - a_o \|_\rho \leq \varepsilon, \ \| b_{e_o} - b_o \|_\rho \leq \varepsilon.$$

Then, if for $0<\theta\leq\theta_\varepsilon$ _and_ $e\geq e_o$ _we have_

$$\| a_e - a_{e_o} \|_\rho \leq L_1 \varepsilon |\rho|^{e_o}$$

$$\| b_e - b_{e_o} \|_\rho \leq L_2 \varepsilon |\rho|^{e_o}$$

it follows that

$$\| \text{red}^{e+1}(a_e b_e - a_{e_o} b_{e_o}) \|_\rho \leq \varepsilon |\rho|^{e_o}$$

$$\| \text{red}^{e+1} a_e b_e \|_\rho \leq 3\varepsilon |\rho|^{e_o},$$

for $0<\theta\leq\theta_\varepsilon$ _and_ $e\geq e_o$, where red^{e+1} _means terms of order_ $e+1$ _only._

Proof: After writing

$$a_e b_e - a_{e_o} b_{e_o} = (a_e - a_{e_o})(b_{e_o} - b_o) + (a_e - a_{e_o}) b_o + a_o (b_e - b_{e_o})$$

$$+ (a_{e_o} - a_o)(b_e - b_{e_o}) + (a_e - a_{e_o})(b_e - b_{e_o})$$

we obtain for the terms of order $e+1$:

$$\text{red}^{e+1}(a_e b_e - a_{e_o} b_{e_o}) = \text{red}^{e+1}(a_e - a_{e_o})(b_{e_o} - b_o) + \text{red}^{e+1}(a_{e_o} - a_o)(b_e - b_{e_o})$$

$$+ \text{red}^{e+1}(a_e - a_{e_o})(b_e - b_{e_o}).$$

From this we obtain the estimate, cf. (5.1), (ii),

$$\| \text{red}^{e+1}(a_e b_e - a_{e_o} b_{e_o}) \|_\rho \leq 2[L_1 \varepsilon |\rho|^{e_o} \varepsilon + L_2 \varepsilon |\rho|^{e_o} \varepsilon + L_1 L_2 \varepsilon^2 |\rho|^{2e_o}]$$

$$\leq \varepsilon |\rho|^{e_o}.$$

Further we can write

$$\mathrm{red}^{e+1} a_e b_e = \mathrm{red}^{e+1}(a_e b_e - a_{e_o} b_{e_o}) + (\mathrm{red}_{e+1} a_{e_o} b_{e_o} - \mathrm{red}_{e_o} a_{e_o} b_{e_o})$$
$$- (\mathrm{red}_e a_{e_o} b_{e_o} - \mathrm{red}_{e_o} a_{e_o} b_{e_o})$$

and from this we obtain the estimate

$$\| \mathrm{red}^{e+1} a_e b_e \|_\rho \leq 3\varepsilon |\rho|^{e_o}$$

for $o<\theta\leq\theta_\varepsilon$ and $e\geq e_o$.

(5.8.1) Let the assumptions be as in (5.8), write

$$a_{e+1} = a_e + \sum_{\nu\in\Delta_{e+1}} a_\nu t^\nu, \quad b_{e+1} = b_e + \sum_{\nu\in\Delta_{e+1}} b_\nu t^\nu \text{ and}$$

assume in addition to the estimates in (5.8) that with a constant $L_o>o$ independent of L_1, L_2 we have

$$\rho^\nu \| a_\nu \| \leq L_o \ \varepsilon |\rho|^{e_o} \gamma_\nu$$
$$\rho^\nu \| b_\nu \| \leq L_o \ \varepsilon |\rho|^{e_o} \gamma_\nu$$

for $\nu\in\Delta_{e+1}$, $o<\theta\leq\theta_\varepsilon$, $e\geq e_o$. Then with $L_o^* = 3 + 2L_oM_o$, where $|a^o|$, $|b^o| \leq M_o$, we have

$$\| \mathrm{red}_{e+1} a_{e+1} b_{e+1} - \mathrm{red}_e a_e b_e \|_\rho \leq L_o^* \ \varepsilon |\rho|^{e_o}.$$

Proof: After writing

$$\mathrm{red}_{e+1} a_{e+1} b_{e+1} - \mathrm{red}_e a_e b_e$$
$$= \mathrm{red}_{e+1}(a_{e+1} b_{e+1} - a_e b_e) + \mathrm{red}_{e+1} a_e b_e - \mathrm{red}_e a_e b_e$$
$$= \mathrm{red}_{e+1}((a_{e+1} - a_e) b_{e+1} + a_e (b_{e+1} - b_e)) + \mathrm{red}^{e+1} a_e b_e$$
$$= (a_{e+1} - a_e) b_o + a_o (b_{e+1} - b_e) + \mathrm{red}^{e+1} a_e b_e$$

we obtain the estimate

$$\| \mathrm{red}_{e+1} a_{e+1} b_{e+1} - \mathrm{red}_e a_e b_e \|_\rho \leq 2L_oM_o \ \varepsilon |\rho|^{e_o} + 3\varepsilon |\rho|^{e_o} = L_o^* \varepsilon |\rho|^{e_o}.$$

§ 6 Norms

From now on we will need norms to construct an analytic versal deformation. We will make use of the ρ-norms defined by a special majorant as introduced in [6], cf. § 5. However for the coefficients of the power series we are going to use L^2-norms instead of L^∞-norms.

(6.1) Let Ω be a domain in $\mathbb{C}^n$. By $H(\Omega)$ we denote the Hilbert-space of square integrable holomorphic functions and for any $f \in H(\Omega)$ we denote its norm by

$$\|f\|_\Omega = \left(\int_\Omega |f|^2 d\lambda\right)^{1/2}$$

where λ is the Lebesgue-measure of $\mathbb{C}^n$. If $f = (f_1,\ldots,f_p)$ is a tuple of functions in $H(\Omega)$ we set

$$\|f\|_\Omega = \max \|f_i\|_\Omega$$

and if $A \in \mathrm{Hom}(\Omega, \mathcal{O}^p, \mathcal{O}^q)$ is a matrix whose entries are in $H(\Omega)$, we define

$$\|A\|_\Omega = \sup\{\|Af\|_\Omega,\ f \in \Gamma(\Omega, \mathcal{O}^p),\ \|f\|_\Omega = 1\}\ .$$

Then for two matrices $A \in \mathrm{Hom}(\Omega, \mathcal{O}^p, \mathcal{O}^q)$ and $B \in \mathrm{Hom}(\Omega, \mathcal{O}^q, \mathcal{O}^r)$ we have $\|BA\|_\Omega \leq \|B\|_\Omega \|A\|_\Omega$.

(6.2) If X is a compact space we define norms for sections in $\mathcal{O}_X$ by employing special charts. A special chart is a tuple $\mathfrak{U} = (U,Y,j,\Omega,D)$ where $Y \subset X$ is an open subset, Ω is an open set in some $\mathbb{C}^n$, $j: Y \to \Omega$ is a closed imbedding of complex spaces, and $D \subset \Omega$ is a Stein open subset with $U = j^{-1}D$. By abuse of notation we will often identify $\mathcal{U}$ and U. Since D is Stein by theorem B we have a canonical epimorphism

$$\Gamma(D, \mathcal{O}_D) \xrightarrow[j^*]{} \Gamma(U, \mathcal{O}_X)$$

and for any $f \in \Gamma(U, \mathcal{O}_X)$ we define

$$\|f\|_U = \|f\|_{\mathcal{U}} = \inf\{\|\tilde{f}\|_D \mid \tilde{f} \in \Gamma(D, \mathcal{O}_D),\ j^*\tilde{f} = f\}.$$

This norm need not be finite, but the subspace of $\Gamma(U, \mathcal{O}_X)$ of all elements with finite norm is a Banach space. If $A \in \mathrm{Hom}(U, \mathcal{O}_X^p, \mathcal{O}_X^q)$ we define $\|A\|_U$ in the same way as in (6.1) for domains in $\mathbb{C}^n$, such that we have the rule $\|BA\|_U \leq \|B\|_U \|A\|_U$.

(6.3) <u>Let</u> $\mathcal{U}_1$ <u>and</u> $\mathcal{U}_2$ <u>be two charts as above and let</u> $U_1 \subset\subset U_2$. <u>Then there exists a constant</u> $K>0$ <u>such that for any</u> $f \in \Gamma(U_2, \mathcal{O}_X)$ <u>we have</u>

$$\|f|U_1\|_{\mathcal{U}_1} \leq K\|f\|_{\mathcal{U}_2}.$$

Proof: Consider the restriction map $\Gamma(U_2, \mathcal{O}_X) \xrightarrow[r]{} \Gamma(U_1, \mathcal{O}_X)$ of the section modules endowed with the canonical Frechet topology. The sets $A_i = \{f \in \Gamma(U_i, \mathcal{O}_X)$ with $\|f\|_{\mathcal{U}_i} \leq 1\}$ are bounded in $\Gamma(U_i, \mathcal{O}_X)$ respectively as can be seen by going back to $\Gamma(D_i, \mathcal{O}_{D_i})$ and using Cauchy's formula. But $r^{-1}A_1$ is a neighborhood of 0 in $\Gamma(U_2, \mathcal{O}_X)$. Hence there is a $K>0$ such that $A_2 \subset K\, r^{-1}A_1$ which implies our assertion.

Remark: Lemma (6.3) is used in § 9, § 10 where we have to make use of several charts. However in § 6, § 7, § 8 we only need norms defined by a single chart. Norms for Čech-cochains defined by several charts will be defined in § 9.

(6.4) Let X be a closed analytic subspace of a domain Ω in $\mathbb{C}^n$. Let $D \subset\subset \Omega$ be a Stein open subset and let (D_i) be a finite Stein open covering of $\bar{D}$. Let $U = X \cap D$ and $\mathfrak{U} = (U_i)$ with $U_i = X \cap D \cap D_i$. Then U and U_i and $U_{i_0 \ldots i_p}$ can be considered as charts as in (6.2) belonging to one and the same imbedding $X \subset \Omega$. For any $\xi \in C^p(\mathfrak{U}, \mathcal{O}_X)$ with components $\xi_{i_0 \ldots i_p} \in \Gamma(U_{i_0 \ldots i_p}, \mathcal{O}_X)$ we define

$$\|\xi\|_{\mathfrak{U}} = \max \|\xi_{i_0 \ldots i_p}\|_{U_{i_0 \ldots i_p}}$$

where $\|\xi_{i_0 \ldots i_p}\|_{U_{i_0 \ldots i_p}}$ is defined as in (6.2). By considering the natural Frechet-topology on $C^p(\mathfrak{U}, \mathcal{O}_X)$ and using the open mapping theorem one easily obtains

(6.4.1) Lemma: Let $X \subset \Omega$ be as above, let $D' \subset\subset D \subset\subset \Omega$ be Stein open subsets, and let (D_i') resp. (D_i) be finite Stein coverings of $\overline{D'}$ resp. $\bar{D}$ such that for any i we have $D_i' \subset\subset D_i$. Let $U' = X \cap D'$, $U = X \cap D$, $U_i' = U' \cap D_i'$, $U_i = U \cap D_i$ and let $\mathfrak{U}' = (U_i')$, $\mathfrak{U} = (U_i)$. Then there is a constant $K > 0$ such that for any $\xi \in Z^s(\mathfrak{U}, \mathcal{O}_X)$, $s > 0$, there exists $\eta \in C^{s-1}(\mathfrak{U}', \mathcal{O}_X)$ such that $\delta\eta = \xi | \mathfrak{U}'$ and $\|\eta\|_{\mathfrak{U}'} \leq K \|\xi\|_{\mathfrak{U}}$.

In the proof of the main Lemmas (7.1), (7.2) we will need an analogous statement without shrinking U resp. D. This can be achieved by choosing D privileged.

(6.5) If $\mathcal{F}$ is a coherent analytic sheaf on a Stein domain Ω in C^n together with a global epimorphism $\mathcal{O}^p \to \mathcal{F}$ such that also $\Gamma(\Omega, \mathcal{O}^p) \xrightarrow[\pi]{} \Gamma(\Omega, \mathcal{F})$ is surjective, then for any $s \in \Gamma(\Omega, \mathcal{F})$ we define

$$\|s\|_\Omega = \inf\{\|f\|_\Omega \mid f \in \Gamma(\Omega, \mathcal{O}^p),\ \pi f = s\}$$

This norm may be infinite, but the subspace $\Gamma_o(\Omega, \mathcal{F})$ of $\Gamma(\Omega, \mathcal{F})$ of the elements with finite norm is a Banach space. Note that the definition of $\|s\|_\Omega$ depends on the representation $\mathcal{O}^p \to \mathcal{F}$, but if $\Omega \subset\subset \mathbb{C}^n$ and we only consider representations of $\mathcal{F}$ in a neighborhood of Ω then $\Gamma_o(\Omega, \mathcal{F})$ will be independent of these representations.

Let now $G = G_1 \times \ldots \times G_n$ be a polydomain in $\mathbb{C}^n$ where each G_i is a bounded convex open set in $\mathbb{C}$. Let $\mathcal{F}$ be coherent in a neighborhood of $\overline{G}$. The polydomain G is called $\mathcal{F}$ -privileged, if for some (and then any) resolution

$$0 \longrightarrow \mathcal{O}^{p_d} \longrightarrow \ldots \longrightarrow \mathcal{O}^{p_o} \longrightarrow \mathcal{F} \longrightarrow 0$$

in a neighborhood of $\overline{G}$ the induced sequence

$$0 \longrightarrow H(G)^{p_d} \longrightarrow \ldots \longrightarrow H(G)^{p_1} \xrightarrow[\alpha]{} H(G)^{p_o}$$

of the Hilbert-spaces in exact and $\operatorname{Im}\alpha$ is closed. In this case $H(G, \mathcal{F}) = \operatorname{Coker} \alpha$ is a Hilbert-space independent of the resolution. Note that G is $\mathcal{F}$-privileged if and only if the canonical maß $H(G, \mathcal{F}) \longrightarrow \Gamma(G, \mathcal{F})$ is injective, [15], [19]. Moreover in this case $H(G, \mathcal{F}) = \Gamma_o(G, \mathcal{F})$.

In order not to shrink special domains by estimating norms we also use the following Lemma, which results from **Hörmander's** L^2 estimates of $\bar{\partial}$, [13], [14].

(6.6) Lemma: Let $\Omega \subset\subset \mathbb{C}^n$ be a Stein domain and let $\mathfrak{U} = (U_i)$ be a finite Stein cover of $\bar{\Omega}$. Then there exists a constant $K>0$ such that the following holds: For any $\xi \in Z^s(\Omega\cap\mathfrak{U}, \mathcal{O})$, $s>0$, there exists $\eta \in C^{s-1}(\Omega\cap\mathfrak{U}, \mathcal{O})$ such that $\delta\eta = \xi$ and $\|\eta\|_{\Omega\cap\mathfrak{U}} \leq K\|\xi\|_{\Omega\cap\mathfrak{U}}$.

This Lemma can be proved by the following result of Hörmander, [14], th. (2.2.3). Denote by $L_{p,q}(\Omega)$ the space of (p,q)-forms $u = \Sigma\, u_{IJ}\, dz_I \wedge d\bar{z}_J$ with coefficients $u_{IJ} \in L^2(\Omega)$ and with norm

$$\|u\|^2 = \sum_{I,J} \int_\Omega |u_{IJ}|^2 d\lambda .$$

Then there exists a constant $K>0$ such that for any $f \in L_{p,q}(\Omega)$, $q>0$, $\bar{\partial}f = 0$ there exists $g \in L_{p,q-1}(\Omega)$ such that $\bar{\partial}g = f$ and $\|g\| \leq K\|f\|$, without shrinking Ω. The proof of (6.6) can now be derived by the usual $\bar{\partial}$-δ-double complex, cf. [13], (7.6.1), without shrinking Ω and $\mathfrak{U}$.

(6.7) Lemma: Let $\mathcal{F}$ be a coherent analytic sheaf on a domain Ω in $\mathbb{C}^n$, let $G \subset\subset \Omega$ be a $\mathcal{F}$-privileged polydomain and let

$$\mathcal{O}^q \xrightarrow[\varphi]{} \mathcal{O}^p \longrightarrow \mathcal{F} \longrightarrow 0$$

be a representation of $\mathcal{F}$ in a neighborhood of $\bar{G}$. If $U' \subset\subset U$ are Stein open subsets and $\mathfrak{U}' = (U'_i)$ resp. $\mathfrak{U} = (U_i)$ with

$U_i' \subset\subset U_i$ are finite Stein coverings of $\overline{G}\cap\overline{U}'$ resp. $\overline{G}\cap\overline{U}$, then there is a constant $K>0$ with the following property:

(i) If $f \in \Gamma(G\cap U, \operatorname{Im}\varphi)$ there exists $g \in \Gamma(G\cap U', \mathcal{O}^q)$ such that $f = \varphi g$ and $\|g\|_{G\cap U'} \leq \|f\|_{G\cap U}$.

(ii) If $\xi \in Z^s(G\cap U\cap \mathfrak{U}, \operatorname{Im}\varphi)$, $s>0$, there exists $\eta \in C^{s-1}(G\cap U'\cap \mathfrak{U}', \operatorname{Im}\varphi)$ such that $\xi | G\cap U'\cap \mathfrak{U}' = \delta\eta$ and $\|\eta\|_{G\cap U'\cap\mathfrak{U}'} \leq K\|\xi\|_{G\cap U\cap\mathfrak{U}}$.

Proof: Since G is privileged we can assume that $\mathcal{F}$ has a resolution

$$0 \longrightarrow \mathcal{O}^{p_m} \longrightarrow \cdots \longrightarrow \mathcal{O}^{p_2} \xrightarrow[\psi]{} \mathcal{O}^q \xrightarrow[\varphi]{} \mathcal{O}^p \rightarrow \mathcal{F} \longrightarrow 0$$

in a neighborhood of $\overline{G}$. Then G is privileged with respect to the cokernels of all the homomorphisms of the resolution. We proceed now by induction on the length of the resolution. The induction basis is given by (6.4). Hence we can assume that the statement of the Lemma is true for ψ.

Let $U' \subset\subset V \subset\subset U$ be an intermediate Stein open subset. Then for any $x \in \overline{G}\cap\overline{V}$ we can find a polydomain V_x such that $V_x \subset\subset U$ and $G\cap V_x$ is again $\mathcal{F}$-privileged. This can be done by the characterization of privileged sets as in [15], [19]. Since $\overline{G}\cap\overline{V}$ is compact we can find a finite Stein-covering $\mathfrak{V} = (V_j)$ of $\overline{G}\cap\overline{V}$ such that $G\cap V_j$ is privileged and $V_j \subset\subset U$.

a) If now $f \in \Gamma(G \cap U, \operatorname{Im}\varphi)$ is given we find $g_i \in \Gamma(G \cap V_i, \mathcal{O}^q)$ such that $\varphi g_i = f | G \cap V_i$ and $\| g_i \|_{G \cap V_i} \leq K_o \| f \|_{G \cap U}$ where K_o is a constant depending only on the geometric situation. Define now $\xi_{ij} = g_j - g_i$. Then

$$\xi = (\xi_{ij}) \in Z^1(G \cap V \cap \mathfrak{V}, \operatorname{Im}\psi) .$$

Let $\mathfrak{V}' = (V_j')$ with $V_j' \subset\subset V_j$ be a Stein covering of $\overline{U} \cap \overline{G}$. Then by induction hypothesis on ψ there exists a constant K_1 and $h \in C^o(G \cap U' \cap \mathfrak{V}', \operatorname{Im}\psi)$ such that

$$\xi_{ij} = h_j - h_i \text{ on } G \cap U' \cap V_{ij}' \text{ and}$$

$$\| h \|_{G \cap U' \cap \mathfrak{V}'} \leq K_1 \| \xi \|_{G \cap V \cap \mathfrak{V}} \leq 2 K_o K_1 \| f \|_{G \cap U} .$$

Then $g = g_i - h_i$ is globally defined on $G \cap U'$ and $\varphi g = f | G \cap U'$, and with some geometric constant K_2 we have

$$\| g \|_{G \cap U'} \leq K_2 \| (g_i - h_i) \|_{G \cap U' \cap \mathfrak{V}'}$$

$$\leq (2 K_o K_1 + K_o) K_2 \| f \|_{G \cap U} .$$

b) Now we prove (ii) by using (i) and (ii) for $\operatorname{Im}\psi$. Let $U' \subset\subset V \subset\subset U$ be an intermediate Stein open set and also $U' \subset\subset V_i \subset\subset U_i$ for any i, such that $\mathfrak{V} = (V_i)$ covers $\overline{G} \cap \overline{V}$.

If $\xi \in Z^s(G \cap U \cap \mathfrak{U}, \operatorname{Im}\varphi)$ is given, we find, by using (i), sections

$$\zeta_{i_o \ldots i_s} \in \Gamma(G \cap V \cap V_{i_o \ldots i_s}, \mathcal{O}^q)$$

defining a cochain ζ such that $\varphi\zeta = \xi | G \cap V \cap \mathfrak{V}$ and

$$\| \zeta \|_{G \cap V \cap \mathfrak{V}} \leq K \| \xi \|_{G \cap U \cap \mathfrak{U}} .$$

Now $\varphi\delta\zeta = \delta\varphi\zeta = \delta\xi = 0$ and hence

$$\delta\zeta \in Z^{s+1}(G\cap V\cap\mathfrak{W}, \operatorname{Im}\psi)$$

with $\|\delta\zeta\|_{G\cap V\cap\mathfrak{W}} \leq (s+2)\ \|\zeta\|_{G\cap V\cap\mathfrak{W}}$.

By the induction hypothesis on ψ we find $\tilde{\eta} \in C^s(G\cap U'\cap\mathfrak{U}', \operatorname{Im}\psi)$ with $\delta\zeta = \delta\tilde{\eta}$ and

$$\|\tilde{\eta}\|_{G\cap U'\cap\mathfrak{U}'} \leq K^2(s+2)\|\xi\|_{G\cap U\cap\mathfrak{U}}.$$

Now for $\zeta - \tilde{\eta} \in Z^s(G\cap U'\cap\mathfrak{U}', \mathcal{O}^q)$ by (6.4) with a constant K' we find $\gamma \in C^{s-1}(G\cap U'\cap\mathfrak{U}', \mathcal{O}^q)$ such that $\zeta - \tilde{\eta} = \delta\gamma$ and

$$\begin{aligned}\|\gamma\|_{G\cap U'\cap\mathfrak{U}'} &\leq K'\|\zeta - \tilde{\eta}\|_{G\cap U'\cap\mathfrak{U}'}\\ &\leq K'(K^2(s+2) + K)\|\xi\|_{G\cap U\cap\mathfrak{U}}.\end{aligned}$$

Now $\xi = \varphi\zeta = \varphi\delta\gamma = \delta\varphi\gamma$ and hence $\eta = \varphi\gamma$ is a solution, for with some other constant C we have $\|\varphi\gamma\|_{G\cap U'\cap\mathfrak{U}'} \leq C\|\gamma\|_{G\cap U\cap\mathfrak{U}}$ since φ is defined and bounded in a neighborhood of $\overline{G}$. Taking the new constant $CK'K(K(s+2) + 1)$ the assertion (ii) is proved.

Remark: Lemma (6.7) can be applied especially for $\mathcal{O}_X$ if X is a closed subspace of Ω . Moreover we obtain the same results if the homomorphism φ is only defined on X.

(6.7.1) Corollary: Let the domains and coverings be as in (6.7) and let X be a closed analytic subspace of Ω and suppose that a representation $\mathcal{O}_X^q \xrightarrow{\varphi} \mathcal{O}_X^p \longrightarrow \mathcal{F} \longrightarrow 0$ on X of a coherent $\mathcal{O}_X$-sheaf $\mathcal{F}$ is given. Let G be privileged for $\mathcal{F}$ and $\mathcal{O}_X$, and let $D = X\cap G$. Then there exists a constant $K>0$ such that the following holds.

(i) <u>For any</u> $f \in \Gamma(D\cap U, \operatorname{Im}\varphi)$ <u>there exists</u> $g \in \Gamma(D\cap U', \mathcal{O}_X^q)$ <u>such that</u> $\varphi g = f$ <u>and</u> $\|g\|_{D\cap U'} \leq K\|f\|_{D\cap U}$.

(ii) <u>For any</u> $\xi \in Z^s(D\cap U\cap \mathfrak{U}, \operatorname{Im}\varphi)$, $s>0$, <u>there exists</u> $\eta \in C^{s-1}(D\cap U'\cap \mathfrak{U}', \operatorname{Im}\varphi)$ <u>such that</u> $\delta\eta = \xi$ on $D\cap U'\cap\mathfrak{U}'$ <u>and</u>

$$\|\eta\|_{D\cap U'\cap\mathfrak{U}'} \leq K\|\xi\|_{D\cap U\cap\mathfrak{U}}.$$

Proof: Let $\mathcal{O}^r \xrightarrow{\chi} \mathcal{O} \xrightarrow{\pi} \mathcal{O}_X \longrightarrow 0$ be a representation of $\mathcal{O}_X$ in a neighborhood of $\overline{G}$. By theorem B there is a homomorphism $\tilde{\varphi}: \mathcal{O}^q \longrightarrow \mathcal{O}^p$ in a neighborhood of $\overline{G}$ inducing φ on X intersected with that neighborhood. Then we obtain a representation of $\mathcal{F}$ as $\mathcal{O}$-module

$$\mathcal{O}^q \oplus \mathcal{O}^{rp} \xrightarrow{\hat{\varphi}} \mathcal{O}^p \xrightarrow{\hat{\pi}} \mathcal{F} \longrightarrow 0$$

where $\hat{\pi}: \mathcal{O}^p \to \mathcal{O}_X^p \to \mathcal{F}$ and $\hat{\varphi} = \tilde{\varphi} \oplus \chi \oplus \ldots \oplus \chi$. To verify (i) if $f \in \Gamma(D\cap U, \operatorname{Im}\varphi)$, any $\hat{f} \in \Gamma(G\cap U, \mathcal{O}^p)$ inducing f is already contained in $\Gamma(G\cap U, \operatorname{Im}\hat{\varphi})$. If $\|f\|_{D\cap U} < \infty$ then there exists $\hat{f}$ with $\pi\hat{f} = f$ and $\|\hat{f}\|_{G\cap U} \leq 2\|f\|_{D\cap U}$. Now applying (6.7) we can solve $\hat{\varphi}\hat{g} = \hat{f}$ with $\|\hat{g}\|_{G\cap U'} \leq 2K\|f\|_{D\cap U}$. If $\hat{g} = \tilde{g} \oplus \tilde{h}$, then $g = \pi\tilde{g}$ is the solution of $\varphi g = f$ with bounds.

To verify (ii) if $\xi \in Z^s(D\cap U\cap\mathfrak{U}, \operatorname{Im}\varphi)$ is given we can find $\tilde{\xi} \in C^s(G\cap U\cap\mathfrak{U}, \mathcal{O}^p)$ with $\pi\tilde{\xi} = \xi$ and $\|\tilde{\xi}\|_{G\cap U\cap\mathfrak{U}} \leq 2\|\xi\|_{D\cap U\cap\mathfrak{U}}$. Then $\pi\delta\tilde{\xi} = 0$ and hence $\delta\tilde{\xi} \in Z^{s+1}(G\cap U\cap\mathfrak{U}, \operatorname{Im}\chi^p)$. Since G is privileged for $\mathcal{O}_X$ we can find by (6.7) $\tilde{\zeta} \in C^s(G\cap V\cap\mathfrak{W}, \operatorname{Im}\chi^p)$ with $\delta\tilde{\zeta} = \delta\tilde{\xi}$ and $\|\tilde{\zeta}\|_{G\cap V\cap\mathfrak{W}} \leq 2K'(s+2)\|\xi\|_{D\cap U\cap\mathfrak{U}}$. Here $U' \subset\subset V \subset\subset U$ and $U_i' \subset\subset V_i \subset\subset U_i$ are intermediate Stein open subsets. Now $\hat{\xi} = \tilde{\xi} - \tilde{\zeta} \in Z^s(G\cap V\cap\mathfrak{W}, \operatorname{Im}\hat{\varphi})$. Again by (6.7) we find $\hat{\eta} \in C^{s-1}(G\cap U'\cap\mathfrak{U}', \operatorname{Im}\varphi)$ such that $\delta\hat{\eta} = \hat{\xi}$ and

$$\|\hat{\eta}\|_{G\cap U'\cap \mathcal{U}'} \leq K\|\hat{\xi}\|_{G\cap V\cap \mathcal{W}}$$

$$\leq 2(KK'(s+2)+1)\|\xi\|_{D\cap U\cap \mathcal{U}} .$$

Restricting $\hat{\eta}$ to $\mathcal{O}_X$ we find the desired solution.

(6.8) Lemma: Let the geometrical situation be as in (6.7) and let G be privileged for $\mathcal{O}_X$ and $\mathcal{F}$. Let $D = X\cap G$ and let epimorphisms $\mathcal{O}_X^p \xrightarrow[\pi]{} \mathcal{F}$ on a neighborhood of $\bar{D}$ and $\mathcal{O}_X^{p_i} \xrightarrow[\pi_i]{} \mathcal{F}$ on $U_i\cap X$ be given. Denote by $\|\ \|^{\pi}$ resp. $\|\ \|^{\pi_i}$ the norms induced by π resp. π_i. Then there exists a constant $K>0$ such that the following holds:

(i) For any $\xi \in Z^o(D\cap U\cap \mathcal{U}, \mathcal{F})$ with corresponding section $s \in \Gamma(D\cap U, \mathcal{F})$

$$\|s\|^{\pi}_{D\cap U'} \leq K\|\xi\|^{(\pi_i)}_{D\cap U\cap \mathcal{U}}$$

where $\|\xi\|^{(\pi_i)}_{D\cap U\cap \mathcal{U}} = \max_i \|\xi_i\|^{\pi_i}_{D\cap U\cap U_i}$.

(ii) For any $\xi \in Z^s(D\cap U\cap \mathcal{U}, \mathcal{F})$, $s>0$, there exists $\eta \in C^{s-1}(D\cap U'\cap \mathcal{U}', \mathcal{F})$ such that $\delta\eta = \xi | D\cap U'\cap \mathcal{U}'$ and

$$\|\eta\|^{\pi}_{D\cap U'\cap \mathcal{U}'} \leq K\|\xi\|^{\pi}_{D\cap U\cap \mathcal{U}} .$$

Proof: We obtain the canonical epimorphisms

$$\tilde{\pi} : \mathcal{O}_\Omega^p \longrightarrow \mathcal{O}_X^p \xrightarrow[\pi]{} \mathcal{F} \quad \text{resp.} \quad \tilde{\pi}_i : \mathcal{O}_\Omega^{p_i} \longrightarrow \mathcal{O}_X^{p_i} \xrightarrow[\pi_i]{} \mathcal{F}$$

on a neighborhoood of $\bar{G}$ resp. on U_i .

Then by theorem B we find homomorphisms φ on a neighborhood of $\bar{G}$ and ψ_i on U_i such that the following diagram

$$\begin{array}{ccccccc}
\mathcal{O}_\Omega^q & \xrightarrow[\varphi]{} & \mathcal{O}_\Omega^p & \xrightarrow[\pi]{} & \mathcal{F} & \longrightarrow & 0 \\
 & & {\scriptstyle\psi_i}\uparrow & & \| & & \\
 & & \mathcal{O}_\Omega^{p_i} & \xrightarrow[\pi_i]{} & \mathcal{F} & \longrightarrow & 0
\end{array}$$

is commutative and exact. By the definition of the norm $\| \ \|^{\pi_i}$ there exists $f_i \in \Gamma(G\cap U\cap U_i, \mathcal{O}_\Omega^{p_i})$ with $\pi_i f_i = \xi_i$ and $\|f_i\|_{G\cap U\cap U_i} \leq 2\|\xi_i\|^{\pi_i}_{D\cap U\cap U_i}$. Since ψ_i is bounded on $G\cap U^1\cap U_i^1$, where $U' \subset\subset U^1 \subset\subset U$ and $U_i' \subset\subset U_i^1 \subset\subset U_i$, we obtain a constant K_1 such that

$$\|\varphi_i f_i\|_{G\cap U^1\cap U_i^1} \leq K_1 \|\xi\|^{(\pi_i)}_{D\cap U\cap \mathfrak{U}}$$

Now by $g_{ij} = \varphi_j f_j - \varphi_i f_i$ a cocycle in $Z^1(G\cap U^1\cap \mathfrak{U}^1, \operatorname{Im}\varphi)$ is defined. By (6.7) we obtain a second constant K_2 such that we can find $(g_i) \in C^o(G\cap U'\cap \mathfrak{U}', \operatorname{Im}\varphi)$ with $g_{ji} = g_j - g_i$ and

$$\|g_i\|_{G\cap U'\cap U_i'} \leq 2K_2 \ K_1\|\xi\|^{(\pi_i)}_{D\cap U\cap \mathfrak{U}}.$$

Then by $f = \varphi_i f_i - g_i$ a global function is defined such that $\tilde{\pi} f = s$ and $\|f\|_{G\cap U'} \leq K_1(1 + 2K_2)\|\xi\|^{(\pi_i)}_{D\cap U\cap \mathfrak{U}}$. This proves (i).

The proof of (ii) follows the same lines, cf. also (6.7),(ii) and will be omitted.

<u>(6.9) ρ-norms</u>. Let X be a complex space and $\mathfrak{U} = (U,Y,j,\Omega,D)$ a chart. By $\mathcal{B}(\mathfrak{U})$ resp. $\mathcal{B}_{p,q}(\mathfrak{U})$ we denote the Banach spaces of elements of $\Gamma(U, \mathcal{O}_X)$ resp. $\operatorname{Hom}(U, \mathcal{O}_X^p, \mathcal{O}_X^q)$ with finite norm $\| \ \|_{\mathfrak{U}}$.

If $G \subset \mathbb{C}^m$ is a polydisc with center 0, then any p×q matrix $A \in \operatorname{Hom}(U\times G, \mathcal{O}^p_{X\times\mathbb{C}^m}, \mathcal{O}^q_{X\times\mathbb{C}^m})$ has a unique expansion

$$A = \sum_{|\nu|\geq 0} A_\nu t^\nu$$

with coefficients $A_\nu \in \operatorname{Hom}(U, \mathcal{O}_X^p, \mathcal{O}_X^q)$. We define

$$\|A\|_{U,\rho} = \|A\|_{\mathfrak{U},\rho} = \sup_\nu \|A_\nu\|_U \frac{\rho^\nu}{\gamma_\nu}$$

where γ_ν are the coefficients of the majorant in (5.1). This norm need not be finite, but the subspace of matrices with finite norm is a Banach space. Note that for a holomorphic $p\times q$ matrix A und a holomorphic $q\times r$ matrix B on $U\times G$ we have

$$\|B\circ A\|_{U,\rho} \le \|B\|_{U,\rho}\|A\|_{U,\rho} \qquad (*)$$

Let now (S,o) be an analytic subgerm of $(\mathbb{C}^m,o)$ and let S be analytic in the polydisc G. Then we obtain a surjection

$$\Gamma(U\times G, \mathcal{O}_{X\times\mathbb{C}^m}) \xrightarrow[\pi]{} \Gamma(U\times S, \mathcal{O}_{X\times S}) .$$

Then for any $f \in \Gamma(U\times S, \mathcal{O}_{X\times S})$ we define

$$\|f\|_{U,\rho} = \|f\|_{U,\rho} = \inf\{\|\tilde{f}\|_{U,\rho} \mid \tilde{f} \in \Gamma(U\times G, \mathcal{O}_{X\times\mathbb{C}^m}), \pi\tilde{f} = f\}$$

and similarly $\|A\|_{U,\rho}$ for a matrix $A \in \mathrm{Hom}(U\times S, \mathcal{O}^p_{X\times S}, \mathcal{O}^q_{X\times S})$. Then again we obtain the product estimate (*) for matrices on $U\times S$.

<u>(6.10) Notation:</u> Let $(S,o) \subset (\mathbb{C}^m,o)$ be an analytic subgerm with ideal $\mathcal{J}$ and reducing system $\Lambda = \Lambda(\mathcal{J})$ and GW-family $(\omega_\mu)_{\mu\in\Lambda}$. Let R_S be the (nonempty) set of all $\rho \in \mathbb{R}^m_+$ such that $\|\omega_\mu - t^\mu\|_\rho \le \frac{1}{2\operatorname{card}\Lambda}\rho^\mu$, cf, (5.5.1), (iv). By S_e we denote the germ of the truncated ideal $\mathcal{J}_e = \mathcal{J} + \mathfrak{m}^{e+1}/\mathfrak{m}^{e+1}$.

By $GL(n, \mathcal{O}_{X\times S})$ we denote the subsheaf of $\mathcal{H}om(\mathcal{O}^n_{X\times S}, \mathcal{O}^n_{X\times S})$ of invertible $n\times n$-matrices. With these notations the following Lemma holds.

<u>(6.11) Cartan's Lemma, cf. [6], (5.7): Let</u> Q_1, Q_2 <u>be amalgamatable open rectangles of</u> $\mathbb{C}^N$ <u>and let</u> X <u>be a closed analytic</u>

subspace of $Q_1 \cup Q_2$ and set $U_i = X \cap Q_i$ and $U = U_1 \cap U_2$. Then there exist $\varepsilon > 0$ and $K \geq 1$ such that for any germ $(S,o) \subset (\mathbb{C}^m, o)$ and any $\rho \in R_S$ the following holds:

If $\Phi \in \Gamma(U \times S, GL(n, \mathcal{O}_{X \times S}))$ with $\|\Phi - 1\|_{U,\rho} \leq \varepsilon$ there exist $\Phi_i \in \Gamma(U_i \times S, GL(n, \mathcal{O}_{X \times S}))$ such that on $U \times S$

$$\Phi = \Phi_1 \Phi_2^{-1} \quad \text{and} \quad \|\Phi_i - 1\|_{U_i,\rho} \leq K \|\Phi - 1\|_{U,\rho} .$$

If in addition $\Phi | U \times S_e = 1$ then Φ_i can be choosen such that $\Phi_i | U_i \times S_e = 1$.

(6.12) Corollary: Let Q_1, Q_2 and Q_1^*, Q_2^* be pairs of open amalgamatable rectangles in $\mathbb{C}^n$ with $Q_i \subset\subset Q_i^*$ and let X be a closed analytic subspace of $Q_1^* \cup Q_2^*$ and set $U_i = X \cap Q_i$, $U_i^* = X \cap Q_i^*$ and $U = U_1 \cap U_2$. Let $\varepsilon > 0$ and $K \geq 1$ be given by (6.11) for the rectangles Q_1, Q_2.

Let further $(S,o) \subset (\mathbb{C}^m, \ )$ be any subgerm, let $\Phi^e \in \Gamma(U \times S_e, GL(n, \mathcal{O}_{X \times S_e}))$ and assume that for $e_o \leq e$ and the truncated matrix $\Phi^{e_o} = \Phi^e | U \times S_{e_o}$ we have already

$$\Phi^{e_o} = (\Phi_1^{e_o})^{-1} \Phi_2^{e_o}$$

with $\Phi_i^{e_o}$ invertible on $U_i^* \times S_{e_o}$, and finally let

$$\|\Phi_i^{e_o}\|_{U_i,\rho}, \ \|(\Phi_i^{e_o})^{-1}\|_{U_i,\rho} \leq L$$

Then for any $\rho \in R_S$ if $\|\Phi^e - \Phi^{e_o}\|_{U,\rho} \leq \frac{\varepsilon}{L^2}$

there exist extensions Φ_i^e of $\Phi_i^{e_o}$ which are invertible on $U_i \times S_e$ such that

$$\Phi_i^e | U_i \times S_{e_o} = \Phi_i^{e_o} | U_i \times S_{e_o}$$

$$\Phi^e = (\Phi_1^e)^{-1} \Phi_2^e$$

$$\| \Phi_i^e - \Phi_i^{e_o} \|_{U_i,\rho} \leq KL^3 \| \Phi^e - \Phi^{e_o} \|_{U,\rho} .$$

Proof: Let $\Psi^e = \Phi_1^{e_o} \Phi^e (\Phi_2^{e_o})^{-1}$ such that $\Psi^e | U \times S_{e_o} = 1$.

Also we have

$$\begin{aligned} \| \Psi^e - 1 \|_{U,\rho} &= \| \Phi_1^{e_o} \Phi^e (\Phi_2^{e_o})^{-1} - \Phi_1^{e_o} \Phi^{e_o} (\Phi_2^{e_o})^{-1} \|_{U,\rho} \\ &\leq \| \Phi_1^{e_o} \|_{U,\rho} \| \Phi^e - \Phi^{e_o} \|_{U,\rho} \| (\Phi_2^{e_o})^{-1} \|_{U,\rho} \\ &\leq \varepsilon . \end{aligned}$$

By (6.11) we can find Ψ_i^e with $\Psi_i^e | U_i \times S_{e_o} = 1$, $\Psi^e = (\Psi_1^e)^{-1} \Psi_2^e$ and $\| \Psi_i^e - \Psi_i^{e_o} \|_{U_i,\rho} \leq K \| \Psi^e - 1 \|_{U,\rho}$. Now set $\Phi_i^e = \Psi_i^e \Phi_i^{e_o}$. Then $\Phi^e = (\Phi_1^e)^{-1} \Phi_2^e$ and

$$\begin{aligned} \| \Phi_i^e - \Phi_i^{e_o} \|_{U_i,\rho} &\leq \| \Psi_i^e - 1 \|_{U_i,\rho} \| \Phi_i^{e_o} \|_{U_i,\rho} \\ &\leq KL \| \Psi^e - 1 \|_{U,\rho} \\ &\leq KL^3 \| \Phi^e - \Phi^{e_o} \|_{U,\rho} . \end{aligned}$$

Remark: The above estimates are independent of $e \geq e_o$.

§ 7 Constructing homomorphisms with bounds.

Notations: For the whole paragraph let $\mathfrak{J}_e$, $e \leq e_1$ with $e_1 \in \mathbb{N}\{\infty\}$, be a sequence of ideals $\mathfrak{J}_e \subset \mathbb{C}_e\langle t\rangle$, $t=(t_1,\ldots,t_m)$, such that $\mathfrak{J}_e|\mathbb{C}_{e-1}\langle t\rangle = \mathfrak{J}_{e-1}$ and such that $\Lambda(\mathfrak{J}_e) = \Lambda(\mathfrak{J}_{e_o}) = \Lambda$ for some e_o and $e_o \leq e \leq e_1$. By $\omega_\mu = t^\mu + a_\mu$, $\mu \in \Lambda$, we denote the GW-family of $\mathfrak{J}_{e_1}$ and we fix some ρ_o such that

$$\| a_\mu \|_{\rho_o} \leq \frac{1}{2\mathrm{card}\Lambda} \rho_o^{\mu} .$$

We use the notation $\rho = \theta\rho_o$ for $0<\theta\leq 1$. We abbreviate $\mathrm{red}_{\mathfrak{J}_e}$ by red_e and use red^e to denote the e-homogeneous part of red_e. By S_e we denote the germ of $\mathfrak{J}_e$ and we write $\mathcal{O}_e$ instead of $\mathcal{O}_{X\times S_e}$. The restriction map $\mathcal{O}_e \to \mathcal{O}_d$ for $e \geq d$ will simply be denoted by $f \to f|S_d$. If $f|S_d = g$ we say that f extends g. We also write $f_o = f|S_o$, which is an element of $\mathcal{O}_X$. If $f \in \Gamma(U\times S_e, \mathcal{O}_e)$ is represented by a series $\sum\limits_\nu f_\nu t^\nu$ with $f_\nu \in \Gamma(U, \mathcal{O}_X)$, which is Λ-e-reduced, we will write $f = \sum\limits_\nu f_\nu t^\nu$ for simplicity since the reduced representative is uniquely determined. Then $f|S_d = \sum\limits_{|\nu|\leq d} f_\nu t^\nu$ is simply the truncation of the series. The same convention will be used for matrices with entries in $\Gamma(U\times S_e, \mathcal{O}_e)$.
If $\mathfrak{U} = (U,Y,j,\Omega,D)$ is a special chart and if $\| f \|_{U,\rho} < \infty$ then we get such a series with $\| f_\nu \|_U < \infty$ by (5.5.1). In this case f is called bounded and reduced, and the same terminology will be used for matrices.

<u>(7.1) Lemma:</u> <u>Let</u> $\mathfrak{U} = (U,Y,j,\Omega,D)$ <u>be a special chart with</u> $D \subset\subset \Omega$ <u>a polydomain and let</u>

$$\varphi_{e_o} : \mathcal{O}_{e_o}^q \to \mathcal{O}_{e_o}^p$$

<u>be a homomorphism defined in a neighborhood of</u> $\overline{U}$, <u>such that</u> φ_{e_o}

is bounded and reduced. Assume further that Coker φ_{e_o} *is* S_{e_o}*-flat and that* D *is privileged for* $\mathcal{O}_X$ *and* Coker φ_o. *Let moreover*

$$g_{e_o} \in \Gamma(U\times S_{e_o}, \mathcal{O}^q_{e_o}) \quad \text{and} \quad f_{e_o} = \varphi_{e_o} g_{e_o}$$

be given and let f_{e_o} *and* g_{e_o} *be also bounded and reduced. Then the following holds.*

For any constant $L_o>0$ *there exists a constant* $L_1>0$ *depending on* L_o, U, φ_o, g_o *with the property that for any* $L\geq L_1$ *there exists* $\varepsilon_1>0$ *depending on* L, *such that for* $0<\varepsilon\leq\varepsilon_1$ *there exists* θ_1 depending on ε and the e_o-data which for $0<\theta\leq\theta_1$ *satisfies the following.*

If φ_e *extends* φ_{e_o} *such that* Coker φ_e *is* S_e*-flat and* φ_e *is bounded and reduced on* $U\times S_e$ *and if* f_e *extends* f_{e_o} *and is bounded and reduced and if*

$$\|\varphi_e - \varphi_{e_o}\|_{U,\rho} \leq L_o\varepsilon|\rho|^{e_o}$$

$$\|f_e - f_{e_o}\|_{U,\rho} \leq L_o\varepsilon|\rho|^{e_o},$$

then there exists $g_e \in \Gamma(U\times S_e, \mathcal{O}^q_e)$, *bounded, reduced and extending* g_{e_o} *such that* $\varphi_e g_e = f_e$ *and* $\|g_e - g_{e_o}\|_{U,\rho}\leq L\varepsilon|\rho|^{e_o}$.

Remarks: 1) The Lemma and the choice of the constants $L_o, L, \varepsilon, \theta$ with the above conditions is independent of $e\geq e_o$.

2) g_e is constructively obtained: If g_d, $e_o\leq d<e$ is already constructed with $\|g_d - g_{e_o}\|_{U,\rho}\leq L\varepsilon|\rho|^{e_o}$ and $\varphi_d g_d = f_d$, there exists g_{d+1} extending g_d such that $\|g_{d+1} - g_{e_o}\|_{U,\rho}\leq L\varepsilon|\rho|^{e_o}$ and

$\varphi_{d+1} g_{d+1} = f_{d+1}$.

3) The condition on the constants in the Lemma is a so-called triangular condition. More precisely we define

(7.2) Definition: Let geometrical data like $\mathfrak{U}$ and e_o-data like φ_{e_o} and g_{e_o} be given. A triangular condition T depending on this e_o-situation consists of three functions

$L_*(x)>0$, $\varepsilon_*(x,y)>0$, $\theta_*(x,y,z)>0$ each depending on the e_o-situation, such that T is defined as the set of all $(L_o,L,\varepsilon,\theta)$ with $L_o>0$, $L\geq L_*(L_o)$, $0<\varepsilon\leq\varepsilon_*(L_o,L)$, $0<\theta\leq\theta_*(L_o,L,\varepsilon)$. The constants of most of our following estimates will satisfy such triangular conditions. If two triangular conditions T_i with defining functions L^i_*, ε^i_*, θ^i_* are given, then $T_1\cap T_2$ is a triangular condition. We moreover define the composition $T_2\circ T_1$ by

$L_*(L_o) = L^2_* L^1_*(L_o)$, $\varepsilon_*(L_o,L) = \inf(\varepsilon^2_*(L^1_*(L_o),L), \varepsilon^1_*(L_o,L^1_*(L_o)))$

and

$\theta_*(L_o,L,\varepsilon) = \inf(\theta^2_*(L^1_*(L_o),L,\varepsilon),\ \theta^1_*(L_o,L^1_*(L_o),\varepsilon))$.

Proof of (7.1): Let K_o be the pullback constant of φ_o and let L_1 be such that

$$K_o(1 + 2L_o + L_oM_o) < L_1,$$

where $M_o = \|g_o\|_U$. Let then for $L\geq L_1$ the constant $\varepsilon_1>0$ be such that

$$K_o(1 + 2L_o + L_oM_o) + 2\varepsilon_1 K_o(L_o + L + L_oL) \leq L ,$$

let $0<\varepsilon\leq\varepsilon_1$ and let $\theta_1>0$ be finally so small that for $0<\theta\leq\theta_1$ the following estimates hold:

$$\|\varphi_{e_o} - \varphi_o\|_{U,\rho} \leq \varepsilon , \quad \|g_{e_o} - g_o\|_{U,\rho} \leq \varepsilon$$

$$\|\mathrm{red}_d(\varphi_{e_o} g_{e_o} - f_{e_o})\|_{U,\rho} \leq \varepsilon|\rho|^{e_o} \text{ for } d \geq e_o, \text{ cf. (5.7).}$$

b) Let $e_o \leq d < e$ and assume g_d is already constructed such that $\varphi_d g_d = f_d$ on $U \times S_d$ and such that

$$\|g_d - g_{e_o}\|_{U,\rho} \leq L\varepsilon|\rho|^{e_o} ,$$

and assume further that g_d is reduced. We want to find g_ν on U such that

$$(\varphi_d + \sum_{\nu \in \Delta_{d+1}} \varphi_\nu t^\nu)(g_d + \sum_{\nu \in \Delta_{d+1}} g_\nu t^\nu) = f_d + \sum_{\nu \in \Delta_{d+1}} f_\nu t^\nu \bmod \tilde{\mathfrak{J}}_{d+1}.$$

This is equivalent to

$$\sum_{\nu \in \Delta_{d+1}} t^\nu(\varphi_\nu g_o + \varphi_o g_\nu - f_\nu) = \mathrm{red}_{d+1}(f_d - \varphi_d g_d) ,$$

and if we write

$$\mathrm{red}_{d+1}(\varphi_d g_d - f_d) = \sum_{\nu \in \Delta_{d+1}} a_\nu t^\nu ,$$

we obtain the equivalent conditions

$$\varphi_o g_\nu = f_\nu - a_\nu - \varphi_\nu g_o \quad \text{for } \nu \in \Delta_{d+1} .$$

c) We next show that indeed $f_\nu - a_\nu - \varphi_\nu g_o \in \Gamma(U, \mathrm{Im}\ \varphi_o)$. First there exists a homomorphism ψ_e on $U \times S_e$, not necessarily with bounds, such that

$$\mathcal{O}_e^r \xrightarrow[\psi_e]{} \mathcal{O}_e^q \xrightarrow[\varphi_e]{} \mathcal{O}_e^p$$

is exact. Since $f_e \in \mathrm{Im}\ \varphi_e$ we have $f_{d+1} = \varphi_{d+1}\tilde{g}_{d+1}$, where $\tilde{g}_{d+1}$ may not be bounded on U but $\|\tilde{g}_{d+1}\|_{V,\rho} < \infty$ for some $V \subset\subset U$. Let $\tilde{g}_{d+1}$ be reduced with respect to the Banach space

$\mathcal{B}_q(V)$, cf. (6.9), and write $\tilde{g}_{d+1} = \tilde{g}_d + \tilde{g}^{d+1}$ with $\tilde{g}^{d+1} = \sum_{\nu\in\Delta_{d+1}} g_\nu t^\nu$. Because $\varphi_d(\tilde{g}_d - g_d) = 0$ there exists h_d such that

$$\tilde{g}_d - g_d = \psi_d h_d \equiv \psi_{d+1} h_d \mod \tilde{\mathfrak{J}}_d .$$

Since $\tilde{\mathfrak{J}}_d \subset \tilde{\mathfrak{J}}_{d+1}$ we obtain on $V\times S_{d+1}$

$$(\varphi_{d+1} - \varphi_o)(\tilde{g}_d - g_d) \equiv (\varphi_{d+1} - \varphi_o)\psi_{d+1} h_d \mod \tilde{\mathfrak{J}}_{d+1}$$

and hence mod $\tilde{\mathfrak{J}}_{d+1}$

$$\begin{aligned} f_{d+1} - \varphi_{d+1} g_d &\equiv \varphi_{d+1}(\tilde{g}^{d+1} + \tilde{g}_d - g_d) \\ &\equiv \varphi_o \tilde{g}^{d+1} + (\varphi_{d+1} - \varphi_o)(\tilde{g}_d - g_d) + \varphi_o(\tilde{g}_d - g_d) \\ &\equiv \varphi_o(\tilde{g}^{d+1} - \psi_{d+1}h_d + \tilde{g}_d - g_d) . \end{aligned}$$

Now by applying red_{d+1} we obtain

$$\begin{aligned} &\varphi_o \, \mathrm{red}_{d+1}(\tilde{g}_{d+1} - g_d - \psi_{d+1}h_d) \\ = \;& \mathrm{red}_{d+1}(f_{d+1} - \varphi_{d+1}g_d) \\ = \;& \sum_{\nu\in\Delta_{d+1}} f_\nu t^\nu + \mathrm{red}_{d+1}(f_d - \varphi_d g_d) - \sum_{\nu\in\Delta} \varphi_\nu g_o t^\nu \\ = \;& \sum_{\nu\in\Delta_{d+1}} t^\nu (f_\nu - a_\nu - \varphi_\nu g_o) , \end{aligned}$$

and hence we have

$$(f_\nu - a_\nu - \varphi_\nu g_o)|V \in \Gamma(V, \mathrm{Im}\, \varphi_o) .$$

Since U is privileged, cf. (6.7.1), we find $g_\nu \in \Gamma(U, \mathcal{O}_X^q)$ such that

$$\varphi_o g_\nu = f_\nu - a_\nu - \varphi_\nu g_o$$

$$\|g_\nu\|_U \leq K_o\|f_\nu - a_\nu - \varphi_\nu g_o\|_U .$$

d) Now we estimate $\|a_\nu\|_U$ as follows: We write

$$\varphi_d g_d - \varphi_{e_o} g_{e_o} = (\varphi_d-\varphi_{e_o})(g_{e_o}-g_o) + (\varphi_d-\varphi_{e_o})g_o + \varphi_o(g_d-g_{e_o})$$
$$+ (\varphi_{e_o}-\varphi_o)(g_d-g_{e_o}) + (\varphi_d-\varphi_{e_o})(g_d-g_{e_o})$$

and obtain for the terms of order d+1

$$\|\mathrm{red}^{d+1}(\varphi_d g_d-\varphi_{e_o}g_{e_o})\|_{U,\rho} \leq 2[L_o\varepsilon^2|\rho|^{e_o}+L\varepsilon^2|\rho|^{e_o}+L_oL\varepsilon^2|\rho|^{2e_o}]$$
$$\leq 2\varepsilon^2|\rho|^{e_o}(L_o+L+L_oL) .$$

Now by

$$\|\mathrm{red}_{d+1}(\varphi_d g_d - f_d)\|_{U,\rho} \leq \|\mathrm{red}^{d+1}(\varphi_d g_d - \varphi_{e_o}g_{e_o})\|_{U,\rho}$$
$$+ \|\mathrm{red}^{d+1}(\varphi_{e_o}g_{e_o}-f_{e_o})\|_{U,\rho} + \|f_d-f_{e_o}\|_{U,\rho},$$

we obtain

$$\rho^\nu\|a_\nu\|_U \leq \varepsilon|\rho|^{e_o}[2\varepsilon(L_o + L + L_oL) + 1 + L_o]\gamma_\nu .$$

Because $\rho^\nu\|f_\nu\|_U \leq L_o\varepsilon|\rho|^{e_o}\gamma_\nu$ and $\rho^\nu\|\varphi_\nu g_o\|_U \leq L_oM_o\varepsilon|\rho|^{e_o}\gamma_\nu$,

we obtain finally

$$\rho^\nu\|g_\nu\|_U \leq \varepsilon|\rho|^{e_o}K_o[2\varepsilon(L_o+L+L_oL) + 1 + 2L_o + L_oM_o]\gamma_\nu \leq L\varepsilon|\rho|^{e_o}\gamma_\nu .$$

When we define $g_{d+1} = g_d + \sum_{\nu\in\Delta_{d+1}} g_\nu t^\nu$, we do have $\varphi_{d+1}g_{d+1} = f_{d+1}$ on $U\times S_{d+1}$ and the estimate $\| g_{d+1}-g_{e_o}\|_{U,\rho} \leq L\varepsilon|\rho|^{e_o}$.

(7.3) Lemma: Let $\mathfrak{U}$ and φ_{e_o} be as in (7.1) and assume in addition that a second homomorphism ψ_{e_o} is given over $U\times S_{e_o}$ such that the sequence

$$\mathcal{O}^r_{e_o} \xrightarrow[\psi_{e_o}]{} \mathcal{O}^q_{e_o} \xrightarrow[\varphi_{e_o}]{} \mathcal{O}^p_{e_o}$$

is exact and that ψ_{e_o} and φ_{e_o} are bounded und reduced.

Then there exists a triangular condition T, depending on $\mathfrak{U}$, φ_{e_o}, ψ_{e_o}, such that for arbitrary $L_o>0$ and $(L_o,L,\varepsilon,\theta) \in T$ the following holds.

If φ_e extends φ_{e_o} such that Coker φ_e is S_e-flat and φ_e is bounded and reduced and if

$$\|\varphi_e - \varphi_{e_o}\|_{U,\rho} \leq L_o\varepsilon|\rho|^{e_o},$$

then there exists an extension ψ_e of ψ_{e_o} over $U\times S_e$ which is bounded and reduced, such that $\varphi_e\psi_e = 0$ on $U\times S_e$ and

$$\|\psi_e - \psi_{e_o}\|_{U,\rho} \leq L\varepsilon|\rho|^{e_o}.$$

Remark: One has statements similar to the remarks at the end of Lemma (7.1) concerning the independence on $e\geq e_o$ and the constructiveness of ψ_e.

Proof: a) Let φ_{e_o}, ψ_{e_o} be given as in the Lemma and let $M_o = \|\psi_o\|_U$ and let K_o be the pullback constant of φ_o. Let L_o be given and let L_1 satisfy

$$K_o(1 + L_oM_o) < L_1 .$$

If $L\geq L_1$ let $\varepsilon_1>0$ so small that

$$K_o(1 + L_oM_o) + 2\varepsilon_1K_o(L_o + L + L_oL) \leq L ,$$

and let finally for $0<\varepsilon\leq\varepsilon_1$ be Θ_1 so small that for $0<\Theta\leq\Theta_1$ we have

$$\|\varphi_{e_o} - \varphi_o\|_{U,\rho} \leq \varepsilon \ , \ \|\psi_{e_o} - \psi_o\|_{U,\rho} \leq \varepsilon$$

$$\|\mathrm{red}_{d+1}\varphi_{e_o}\psi_{e_o}\|_{U,\rho} \leq \varepsilon|\rho|^{e_o} \text{ for } d\geq e_o \ ,$$

cf. (5.7).

b) If ψ_d is already constructed with $\varphi_d\psi_d = 0$ on $U\times S_d$ and $\|\psi_d - \psi_{e_o}\|_{U,\rho} \leq L\varepsilon|\rho|^{e_o}$, $e_o\leq d<e$, we want to find ψ_ν, $\nu \in \Delta_{d+1}$, such that

$$(\varphi_d + \sum_{\nu\in\Delta_{d+1}} \varphi_\nu t^\nu)(\psi_d + \sum_{\nu\in\Delta_{d+1}} \psi_\nu t^\nu) \equiv 0 \mod \widetilde{\mathfrak{I}}_{d+1} \ .$$

This is equivalent to

$$\varphi_o\psi_\nu = -a_\nu - \varphi_\nu\psi_o \ , \quad \nu \in \Delta_{d+1},$$

where we define the a_ν by $\mathrm{red}_{d+1}(\varphi_d\psi_d) = \sum_{\nu\in\Delta_{d+1}} a_\nu t^\nu$.

c) Let $f_{d+1} := \varphi_{d+1}\psi_d$ as a power series.
Then $f_d := f_{d+1} \mid \mathbb{C}_d\langle t\rangle \equiv 0 \mod \widetilde{\mathfrak{I}}_d$. Hence by the algebraic statement of (7.1) there exists $\tilde{\psi}_\nu$, $\nu \in \Delta_{d+1}$, defined and bounded on U, such that

$$\tilde{\psi}^{d+1} = \sum_{\nu\in\Delta_{d+1}} \tilde{\psi}_\nu t^\nu$$

gives

$$f_{d+1} \equiv - \varphi_{d+1}\tilde{\psi}^{d+1} \mod \widetilde{\mathfrak{I}}_{d+1} \ .$$

If we set $\tilde{\psi}_{d+1} = \psi_d + \tilde{\psi}^{d+1}$, we have a solution, i.e.

$$a_\nu + \varphi_\nu\psi_o \in \Gamma(U, \mathrm{Im}\ \varphi_o).$$

Now since U is privileged we can find ψ_ν with

$$\varphi_o\psi_\nu = - a_\nu - \varphi_\nu\psi_o$$

$$\|\psi_\nu\|_U \leq K_o(\|a_\nu\|_U + M_o\|\varphi_\nu\|_U) .$$

d) To estimate $\|a_\nu\|_U$, we have

$$\|red_{d+1}\varphi_d\psi_d\|_{U,\rho} \leq \|red^{d+1}(\varphi_d\psi_d - \varphi_{e_o}\psi_{e_o})\|_{U,\rho} + \|red^{d+1}\varphi_{e_o}\psi_{e_o}\|_{U,\rho}$$

$$\leq 2\varepsilon^2|\rho|^{e_o}(L_o + L + L_oL) + \varepsilon|\rho|^{e_o}$$

as in (7.1). And again as in (7.1) we obtain finally

$$\|\psi_\nu\|_U \frac{\rho^\nu}{\gamma_\nu} \leq \varepsilon|\rho|^{e_o}K_o[2\varepsilon(L_o + L + L_oL) + 1 + L_oM_o]$$

$$\leq L\ \varepsilon|\rho|^{e_o} .$$

If now $\psi_{d+1} = \psi_d + \sum_{\nu\in\Delta_{d+1}} \psi_\nu t^\nu$, we have

$\varphi_{d+1}\psi_{d+1} = 0$ on $U\times S_{d+1}$ and the estimate $\|\psi_{d+1}-\psi_{e_o}\|_{U,\rho} \leq L\varepsilon|\rho|^{e_o}$.

This completes the proof of (7.3).

We also need an analogue of Lemma (7.1) if U is not privileged. In this case we have to assume that U is Stein and to allow shrinking. In the proof of the following main Lemma (7.4) we use the same method as in the proof of Lemma (6.7) by considering a resolution of finite length. As in (6.7) we have also to consider Cech cocycles. But this time we have to estimate all sections, homomorphisms and cochains by the ρ-norm and to make sure that the estimates are independent of $e \geq e_o$.

<u>(7.4) Main Lemma:</u> <u>Let</u> $\mathfrak{U} = (U,Y,j,\Omega,D)$ <u>be a special chart, let</u> $D' \subset\subset D \subset\subset \Omega$ <u>be Stein such that</u> $U' = j^{-1}D'$ <u>and</u> $U = j^{-1}D$.

Let a homomorphism

$$\varphi_{e_o} : \mathcal{O}^q_{e_o} \longrightarrow \mathcal{O}^p_{e_o}$$

be given over $U \times S_{e_o}$ and assume that φ_{e_o} is bounded and reduced and that $\operatorname{Coker}\varphi_{e_o}$ is S_{e_o}-flat. Then the following conclusions hold

(I) If $g_{e_o} \in \Gamma(U \times S_{e_o}, \mathcal{O}^q_{e_o})$, $f_{e_o} = \varphi_{e_o} g_{e_o}$, and if g_{e_o} and f_{e_o} are bounded and reduced, then there exists a triangular condition T, depending only on U', U, φ_{e_o}, g_{e_o}, such that for any $L_o > 0$ and any (L, ε, θ) with $(L_o, L, \varepsilon, \theta) \in T$ the following is satisfied.

If φ_e resp. f_e extend φ_{e_o} resp. f_{e_o} on $U \times S_e$, $e \geq e_o$, such that both are bounded and reduced as power series, such that, coker φ_e is S_e-flat, and such that

$$\|\varphi_e - \varphi_{e_o}\|_{U,\rho} \leq L_o \varepsilon |\rho|^{e_o}$$

$$\|f_e - f_{e_o}\|_{U,\rho} \leq L_o \varepsilon |\rho|^{e_o} ,$$

then there exists $g_e \in \Gamma(U' \times S_e, \mathcal{O}^q_e)$ bounded, reduced and extending $g_{e_o} | U' \times S_{e_o}$, such that on $U' \times S_e$

$$\varphi_e g_e = f_e \quad \text{and} \quad \|g_e - g_{e_o}\|_{U',\rho} \leq L\varepsilon |\rho|^{e_o} .$$

(II) Let $\mathcal{U}' = (U_i')$, $\mathcal{U} = (U_i)$ with $U_i' \subset\subset U_i$ be finite Stein covers of U' resp. U and let $U_i' = j^{-1} D_i'$, $U_i = j^{-1} D_i$ such that (D_i') resp. (D_i) are finite Stein covers of D' resp. D.

Let $s > 0$ and $\eta_{e_o} \in C^{s-1}(\mathcal{U} \times S_{e_o}, \operatorname{Im} \varphi_{e_o})$ and $\xi_{e_o} = \delta \eta_{e_o}$ be given, such that ξ_{e_o} and η_{e_o} are bounded and reduced as power series with respect to the norm $\| \quad \|_{\mathcal{U},\rho}$, cf. (6.9). Then there exists a triangular condition T such that for any $L_o>0$ and any (L,ε,Θ) with $(L_o,L,\varepsilon,\Theta) \in T$ the following is satisfied.

If φ_e extends φ_{e_o} on $U\times S_e$, such that φ_e is bounded and reduced and such that Coker φ_e is S_e-flat, and if $\xi_e \in Z^s(\mathfrak{U}\times S_e, \operatorname{Im}\varphi_e)$ extends ξ_{e_o} and is bounded and reduced, and if

$$\|\varphi_e - \varphi_{e_o}\|_{U,\rho} \leq L_o\varepsilon|\rho|^{e_o}$$

$$\|\xi_e - \xi_{e_o}\|_{\mathfrak{U},\rho} \leq L_o\varepsilon|\rho|^{e_o}$$

then there exists $\eta_e \in C^{s-1}(\mathfrak{U}'\times S_e, \operatorname{Im}\varphi_e)$ extending η_{e_o}, bounded and reduced, such that $\delta\eta_e = \xi_e|\mathfrak{U}'\times S_e$ and

$$\|\eta_e - \eta_{e_o}\|_{\mathfrak{U}',\rho} \leq L\varepsilon|\rho|^{e_o} .$$

Remarks: 1) As in (7.1), (7.3) the estimates and the choice of the constants are independent of $e\geq e_o$.

2) (I) and (II) are linked by their proof.

3) g_e in (I) and η_e in (II) can be obtained constructively.

Proof: a) Since U is Stein, we can find an Oka-Weil domain W^* such that $U' \subset\subset W^* \subset\subset U$ and such that W^* is a closed analytic subspace of an open rectangle R^* in some $\mathbb{C}^N$. Let $R \subset\subset R^*$ be a second rectangle and let $W = W^*\cap R$ such that $U' \subset\subset W \subset\subset W^*$. Let moreover $\mathfrak{V}' = (V_i')$ resp. $\mathfrak{V} = (V_i)$ be finite Stein coverings of $\overline{R'}$ resp. $\overline{R}$ in R^* such that $V_i' \subset\subset V_i$, $U_i' \subset V_i' \cap U$, $V_i \cap W \subset\subset U_i$. By (6.3) we can assume that all norms occuring in the Lemma are defined via the chart given by $R \subset\subset R^*$ and via the covering $\mathfrak{V}$.

b) Let φ_e be given as in the Lemma and let $\varphi_e = \Sigma\, \varphi_\nu t^\nu$, summation over $\nu \in \Delta$, $|\nu| \leq e$, $\|\varphi_\nu\|_W < \infty$, and such that $\|\varphi_e - \varphi_{e_o}\|_{W,\rho} \leq L_o \varepsilon |\rho|^{e_o}$, i.e. for $e_o < |\nu| \leq e$ we have $\|\varphi_\nu\|_W \rho^\nu \leq L_o \varepsilon |\rho|^{e_o} \gamma_\nu$. Then we can find $\tilde{\varphi}_\nu$ defined on R and inducing φ_ν on W, such that $\|\tilde{\varphi}_\nu\|_R \leq 2\|\varphi_\nu\|_W$. Then with $\tilde{\varphi}_e = \Sigma\, \tilde{\varphi}_\nu t^\nu$ we obtain $\|\tilde{\varphi}_e - \tilde{\varphi}_{e_o}\|_{R,\rho} \leq 2L_o \varepsilon |\rho|^{e_o}$ and $\tilde{\varphi}_e$ induces φ_e on $W \times S_e$.

Let $\mathcal{O}_R^r \xrightarrow{\chi} \mathcal{O}_R$ be a representation of the ideal of W. Putting $\hat{\varphi}_e = \tilde{\varphi}_e \oplus \chi \oplus \ldots \oplus \chi$ we obtain as in (6.8) a homomorphism

$$\hat{\varphi}_e : \mathcal{O}_{R\times S_e}^{q+rp} \longrightarrow \mathcal{O}_{R\times S_e}^{p}$$

such that Coker $\hat{\varphi}_e$ = Coker φ_e, $\hat{\varphi}_e$ is bounded and reduced and

$$\|\hat{\varphi}_e - \hat{\varphi}_{e_o}\|_{R,\rho} = \|\tilde{\varphi}_e - \tilde{\varphi}_{e_o}\|_{R,\rho} \leq 2L_o \varepsilon |\rho|^{e_o} .$$

In the same way as for $\tilde{\varphi}_e$ we find $\hat{f}_e \in \Gamma(R \times S_e, \mathcal{O}_{R\times S_e}^p)$ inducing f_e on $W \times S_e$ such that $\|\hat{f}_e - \hat{f}_{e_o}\|_{R,\rho} \leq 2L_o \varepsilon |\rho|^{e_o}$ if we have $\|f_e - f_{e_o}\|_{W,\rho} \leq L_o \varepsilon |\rho|^{e_o}$.

If finally $\xi_e = \Sigma\, \xi_\nu t^\nu$ is given in $Z^s(\mathfrak{U} \times S_e, \mathcal{O}_{X\times S_e}^p)$, bounded and reduced, we can find $\zeta_\nu \in C^{s-1}(\mathfrak{W} \cap W, \mathcal{O}_X^p)$ with $\delta\zeta_\nu = \zeta_\nu | \mathfrak{W} \cap W$ and $\|\zeta_\nu\|_{\mathfrak{W}\cap W} \leq C\|\xi_\nu\|_{\mathfrak{U}}$, where C is some geometrical constant. Then as before we can find $\hat{\zeta}_\nu \in C^{s-1}(\mathfrak{W}, \mathcal{O}_R^p)$ such that $\|\hat{\zeta}_\nu\|_{\mathfrak{W}} \leq 2\|\zeta_\nu\|_{\mathfrak{W}\cap W}$. Defining now $\hat{\zeta}_e = \Sigma\, \hat{\zeta}_\nu t^\nu$ we obtain a chain in $C^{s-1}(\mathfrak{W} \times S_e, \mathcal{O}_{R\times S_e}^p)$ such that $\delta\hat{\zeta}_e = \hat{\xi}_e$ induces $\xi_e | \mathfrak{W} \cap W \times S_e$. Moreover if $\|\xi_e - \xi_{e_o}\|_{\mathfrak{U},\rho} \leq L_o \varepsilon |\rho|^{e_o}$ we obtain by the above estimates $\|\hat{\xi}_e - \hat{\xi}_{e_o}\|_{\mathfrak{W},\rho} \leq C^* L_o \varepsilon |\rho|^{e_o}$ with some new geometrical constant. Since Coker $\hat{\varphi}_e$ = Coker φ_e, if ξ_e is a cocycle in Im φ_e, then also $\hat{\xi}_e$ is a cocycle in Im $\hat{\varphi}_e$.

c) By a) and b) we can assume without lost of generality that $U = R$ is a rectangle in $\mathbb{C}^N$ and that all norms are defined directly by sets in $\mathbb{C}^N$. We now prove (I) and (II) under the additional assumption that on $R \times S_e$ there exists a global resolution

$$0 \longrightarrow \mathcal{O}_e^{p_m} \xrightarrow[\varphi_e^m]{} \cdots \longrightarrow \mathcal{O}_e^{p_2} \xrightarrow[\varphi_e^2]{} \mathcal{O}_e^q \xrightarrow[\varphi_e]{} \mathcal{O}_e^p$$

with the following property: There exists a triangular condition T depending on the e_o-situation, such that for any $L_o > 0$ and $(L_o^*, \varepsilon, \theta)$, with $(L_o, L_o^*, \varepsilon, \theta) \in T$ and any extension φ_e of φ_{e_o} as in the Lemma, satisfying $\|\varphi_e - \varphi_{e_o}\|_{R,\rho} \leq L_o \varepsilon |\rho|^{e_o}$, we can find φ_e^μ defining a resolution of Im φ_e and such that $\|\varphi_e^\mu - \varphi_{e_o}^\mu\|_{R,\rho} \leq L_o^* \varepsilon |\rho|^{e_o}$.

d) We proceed by induction on m to prove (I), (II). We assume that (I) and (II) have been shown for the homomorphism φ_e^2 and first prove (I). For this purpose let $R' \subset\subset R^1 \subset\subset R$ be an intermediate rectangle and let $\mathfrak{W} = (V_i)$ be a finite covering of $\overline{R}^1$ by privileged polydomains for Coker φ_o, $V_i \subset\subset R$. If $f_{e_o} = \varphi_{e_o} g_{e_o}$ is given, by (7.1) we find triangular conditions T_i such that for $(L_o, L_1, \varepsilon, \theta) \in \cap T_i$ the following holds: If φ_e and f_e extend φ_{e_o} and f_{e_o} respectively with

$$\|\varphi_e - \varphi_{e_o}\|_{R,\rho} \leq L_o \varepsilon |\rho|^{e_o}, \quad \|f_e - f_{e_o}\|_{R,\rho} \leq L_o \varepsilon |\rho|^{e_o}$$

we find $g_{ie} \in \Gamma(V_i \times S_e, \mathcal{O}_e^q)$ extending $g_{e_o}|V_i \times S_{e_o}$, such that $\varphi_e g_{ie} = f_e|V_i \times S_e$ and $\|g_{ie} - g_{e_o}\|_{V_i,\rho} \leq L_1 \varepsilon |\rho|^{e_o}$. Now define $\xi_e = (\xi_{ije}) = (g_{je} - g_{ie})$ on $\mathfrak{V} \cap R^1 \times S_e$. Then $\xi_e|S_{e_o} = 0$, $\xi_e \in Z^1(R^1 \cap \mathfrak{V} \times S_e, \operatorname{Im} \varphi_e^2)$ and $\|\xi_e\|_{\mathfrak{V} \cap R^1 \times S_e} \leq 2L_1 \varepsilon |\rho|^{e_o}$.

Now by induction hypothesis (II) for φ_e^2, if $\mathfrak{V}' = (V_i')$ is finite Stein covering of $\overline{R'}$ with $V_i' \subset\subset V_i$, there exist a triangular condition T' such that for $(L_1, L_2, \varepsilon, \Theta)$ the following holds: There exists $\eta_e \in C^o(R' \cap \mathfrak{V}' \times S_e, \operatorname{Im} \varphi_e^2)$ with $\eta^e|S_{e_o} = 0$, $\delta\eta_e = \xi_e|R' \cap \mathfrak{V}' \times S_e$ and $\|\eta_e\|_{R' \cap \mathfrak{V}', \rho} \leq L_2 \varepsilon |\rho|^{e_o}$.

Putting $g_e = g_{ie} - \eta_{ie}$ on $R' \cap V_i'$ we obtain a globally defined $g_e \in \Gamma(R' \times S_e, \mathcal{O}_e^q)$, bounded and reduced, such that $\varphi_e g_e = f_e$ on $R' \times S_e$ and $\|g_e - g_{e_o}\|_{R', \rho} \leq (L_2 + 2L_1) \varepsilon |\rho|^{e_o}$.
Taking the intersection of all the above triangular conditions, we have proved (I) under the assumption c), since for m = 0 there is nothing to prove.

e) We assume that (I) and (II) are true for φ_e^2 and now prove (II) for φ_e. For this let

$$R' \subset\subset Q' \subset\subset Q \subset\subset R$$
$$\mathfrak{U}' \subset\subset \mathfrak{V}' \subset\subset \mathfrak{V} \subset\subset \mathfrak{U}$$

be intermediate rectangles and Stein coverings with $U_i' \subset\subset V_i' \subset\subset V_i \subset\subset U_i$ and $\mathfrak{V}'$ resp. $\mathfrak{V}$ covering $\overline{Q'}$ resp. $\overline{Q}$.

Let $\xi_{e_o} = \delta\eta_{e_o}$ be given as in the Lemma, assume that $\eta_{e_o} = \varphi_{e_o}\gamma_{e_o}$ with $\gamma_{e_o} \in C^{s-1}(\mathfrak{U}\times S_{e_o}, \mathcal{O}^q_{e_o})$ and let $\zeta_{e_o} = \delta\gamma_{e_o}$, such that $\xi_{e_o} = \varphi_{e_o}\zeta_{e_o}$. We can assume that all these elements are bounded and reduced.

By (I) we find triangular conditions $T_{i_o\ldots i_s}$ depending only on the e_o-situation, such that for $(L_o, L_1, \varepsilon, \theta) \in \cap T_{i_o\ldots i_s}$ the following holds. If φ_e extends φ_{e_o} as above and if $\xi_e \in Z^s(\mathfrak{U}\times S_e, \operatorname{Im}\varphi_e)$ extends ξ_{e_o} such that

$$\|\varphi_e - \varphi_{e_o}\|_{R,\rho} \le L_o\varepsilon|\rho|^{e_o}, \quad \|\xi_e - \xi_{e_o}\|_{\mathfrak{U},\rho} \le L_o\varepsilon|\rho|^{e_o}$$

there exists $\zeta_e \in C^s(\mathfrak{W}\cap Q\times S_e, \mathcal{O}^q_e)$ extending ζ_{e_o} such that $\varphi_e\zeta_e = \xi_e|\mathfrak{W}\cap Q$ and $\|\zeta_e - \zeta_{e_o}\|_{\mathfrak{W}\cap Q,\rho} \le L_1\varepsilon|\rho|^{e_o}$.

Now $\delta\zeta_e \in Z^{s+1}(\mathfrak{W}\cap Q\times S_e, \operatorname{Im}\varphi_e^2)$ and $\delta\zeta_e|S_{e_o} = 0$. By induction hypothesis on (II) there is a triangular condition T_1 only depending on the e_o-situation, such that for $(L_2, \varepsilon, \theta)$ with $(L_1, L_2, \varepsilon, \theta) \in T_1$ we can find

$\eta_e^2 \in C^s(\mathfrak{W}'\cap Q'\times S_e, \operatorname{Im}\varphi_e^2)$ with $\eta_e^2|S_{e_o} = 0$,

$\delta\eta_e^2 = \delta\zeta_e$ on $\mathfrak{W}'\cap Q'$ and the estimate

$$\|\eta_e^2\|_{Q'\cap\mathfrak{W}',\rho} \le L_2\varepsilon|\rho|^{e_o}.$$

Now $\zeta_e - \eta_e^2 \in Z^s(Q'\cap\mathfrak{W}'\times S_e, \mathcal{O}^q_e)$, and since we only deal with $\mathcal{O}_e$ instead of $\operatorname{Im}\varphi_e$, we can find, cf. (6.6),

$\gamma_e \in C^{s-1}(\mathfrak{U}'\times S_e, \mathcal{O}^q_e)$ such that

$$\zeta_e - \eta_e^2 = \delta\gamma_e \quad \text{and} \quad \|\gamma_e - \gamma_{e_o}\|_{\mathfrak{U}',\rho} \leq C(L_1+L_2)\ \varepsilon|\rho|^{e_o}$$

where C is a geometrical constant. (Note that

$\zeta_{e_o} - \eta_{e_o}^2 = \zeta_{e_o} = \delta\gamma_{e_o}$.)

Finally we define $\eta_e = \mathrm{red}_e \varphi_e \gamma_e$ and obtain on $\mathfrak{U}'\times S_e$ that

$$\eta_e = \varphi_e\gamma_e\ ,\ \eta_e|S_e = \eta_{e_o} \quad \text{and}$$

$$\delta\eta_e = \delta\varphi_e\gamma_e = \varphi_e\delta\gamma_e = \varphi_e(\zeta_e - \eta_e^2) = \varphi_e\zeta_e = \xi_e\ .$$

To estimate η_e we obtain

$$\|\eta_e-\eta_{e_o}\|_{\mathfrak{U}',\rho} \leq \|\mathrm{red}_e(\varphi_e\gamma_e-\varphi_{e_o}\gamma_{e_o})\|_{\mathfrak{U}',\rho} + \|\mathrm{red}_e\varphi_{e_o}\gamma_{e_o} - \mathrm{red}_{e_o}\varphi_{e_o}\gamma_{e_o}\|_{\mathfrak{U}',\rho}.$$

The first term can be estimated as in (7.1), d) and the second term can be made $\leq \varepsilon|\rho|^{e_o}$ if the triangular condition is improved, cf. (5.7.1). Thus, taking for T the intersection or composition of all the above triangular conditions we obtain:
If $L_o>0$ and $(L_o,L,\varepsilon,\Theta) \in T$ and if φ_e, ξ_e are given as in the Lemma with

$$\|\varphi_e - \varphi_{e_o}\|_{R,\rho} \leq L_o\varepsilon|\rho|^{e_o} \quad \text{and} \quad \|\xi_e - \xi_{e_o}\|_{\mathfrak{U},\rho} \leq L_o\varepsilon|\rho|^{e_o}\ ,$$

we find η_e with $\|\eta_e - \eta_{e_o}\|_{\mathfrak{U}',\rho} \leq L\varepsilon|\rho|^{e_o}$ such that
$\delta\eta_e = \xi_e|\mathfrak{U}'\times S_e$ and η_e extends η_{e_o} .

f) Now we are going to verify that resolutions as in c) always exist. In order not to use too many notations we will abbreviate the following proof by not specifying the single shrinking. This is justified by the fact that shrinking is only accompanied by multiplying the constants L with geometrical constants and then adding up, a process which only modifies the way the triangular conditions are constructed. First by allowing shrinking we can assume that φ_{e_o} is defined on $R^* \times S_{e_o}$, $R \subset\subset R^*$, R^* a rectangle, and that φ_{e_o} is bounded and reduced. Let (D_i) be a finite covering of $\overline{R}$ by privileged polydomains $D_i \subset\subset R^*$ and let $\{R_j\}$ be a finite family of rectangles contained in R and parallel to R such this family is an open cover of $\overline{R'}$ and such that by successive amalgamations a rectangle can built up which contains $\overline{R'}$. We can assume that each R_j is relatively compact in some D_i. We can assume moreover that on $D_i \times S_{e_o}$ resp. $D_{ij} \times S_{e_o}$ we are given already homomorphisms $\varphi^{\mu}_{ie_o}$ resp. $\psi^{\mu}_{ije_o}$ as in the following commutative diagram

$$\begin{array}{ccccccccccc}
0 & \longrightarrow & \mathcal{O}^{p_m}_{e_o} & \xrightarrow[\varphi^m_{je_o}]{} \cdots \longrightarrow & \mathcal{O}^{p_2}_{e_o} & \xrightarrow[\varphi^2_{je_o}]{} & \mathcal{O}^{q}_{e_o} & \xrightarrow[\varphi_{e_o}]{} & \mathcal{O}^{p}_{e_o} \\
 & & \downarrow{\scriptstyle \psi^m_{ije_o}} & & \downarrow{\scriptstyle \psi^2_{ije_o}} & & \| & & \| \\
0 & \longrightarrow & \mathcal{O}^{p_m}_{e_o} & \xrightarrow[\varphi^m_{ie_o}]{} \cdots \longrightarrow & \mathcal{O}^{p_2}_{e_o} & \xrightarrow[\varphi^2_{ie_o}]{} & \mathcal{O}^{q}_{e_o} & \xrightarrow[\varphi_{e_o}]{} & \mathcal{O}^{p}_{e_o}
\end{array},$$

and such in addition homomorphisms $f^{\mu}_{ije_o}$ on $D_{ij} \times S_{e_o}$ are given with the homotopy identities

$$\mathrm{id} - \psi^{\mu}_{ije_o}\psi^{\mu}_{jie_o} = \varphi^{\mu+1}_{je_o} f^{\mu}_{ije_o} + f^{\mu-1}_{ije_o}\varphi^{\mu}_{je_o} .$$

Furthermore we can assume that all these homomorphisms are bounded and reduced.

g) By (7.3) we find triangular conditions T_i such that for $L_o > 0$ and $(L_o, L_1, \varepsilon, \theta) \in \cap T_i$ the following holds.

If φ_e extends φ_{e_o}, such that Coker φ_e is S_e-flat, φ_e is bounded and reduced with $\|\varphi_e - \varphi_{e_o}\|_{R^*,\rho} \leq L_o\varepsilon|\rho|^{e_o}$, we can find (by finite induction on μ) homomorphisms φ^{μ}_{ie} extending $\varphi^{\mu}_{ie_o}$ on $D_i \times S_e$ such that $\varphi^{\mu+1}_{ie}\varphi^{\mu}_{ie} = 0$ on $D_i \times S_e$ and

$$\|\varphi^{\mu}_{ie} - \varphi^{\mu}_{ie_o}\|_{D_i,\rho} \leq L_1\varepsilon|\rho|^{e_o} . \tag{1}$$

In the same way applying now (I) (in the case where a global resolution is given) to

$$h^{\mu}_{ije_o} = \psi^{\mu-1}_{ije_o}\varphi^{\mu}_{je_o} = \varphi^{\mu}_{ie_o}\psi^{\mu}_{ije_o} ,$$

we find triangular conditions T_{ij} such that for $(L_1,L_2,\varepsilon,\Theta) \in \cap T_{ij}$ and some shrinkings $D'_i \subset\subset D_i$ we have: There exist ψ^{μ}_{ije} extending $\psi^{\mu}_{ije_o}$, bounded and reduced such that $\varphi^{\mu}_{ie}\psi^{\mu}_{ije} = \psi^{\mu-1}_{ije}\varphi^{\mu}_{je}$ on $D'_{ij} \times S_e$ and such that (1) implies

$$\|\psi^{\mu}_{ije} - \psi^{\mu}_{ije_o}\|_{D'_{ij}\rho} \leq L_2\varepsilon|\rho|^{e_o} . \tag{2}$$

This is done inductively over μ where a finite number of shrinkings has to be considered. Finally using (I) again for the equations

$$id - \psi^{\mu}_{jie}\psi^{\mu}_{ije} - f^{\mu-1}_{ije}\varphi^{\mu}_{je} = \varphi^{\mu+1}_{je}f^{\mu}_{ije} \tag{3}$$

on $D'_{ij} \times S_e$, we find triangular conditions T'_{ij} such that for $(L_2,L_3,\varepsilon,\Theta) \in \cap T'_{ij}$ and some shrinkings $D''_i \subset\subset D'_i$ we can find f^{μ}_{ije} on $D''_{ij} \times S_e$, bounded and reduced, such that (3) is satisfied and such that by (2) the estimate

$$\|f^{\mu}_{ije} - f^{\mu}_{ije_o}\|_{D''_{ij}\rho} \leq L_3\varepsilon|\rho|^{e_o} \tag{4}$$

is satisfied.

h) By restricting all homomorphisms φ_e, ψ_e, f_e constructed in g) to the rectangles R_i of the subdivision, we can assume now that all identities and estimates of g) are satisfied with respect to $R_i \times S_e$ and $R_{ij} \times S_e$ and that the indices of φ^μ_{ie}, ψ^μ_{ije}, f^μ_{ije} are now with respect to $\{R_i\}$.

Let R_i and R_j now be amalgamatable. By the algebraic process of amalgamation II in (4.3) we find by elementary algebraic operations commutative diagrams

$$\begin{array}{ccc} \mathcal{O}_e^{\hat{p}_\mu} & \xrightarrow{\hat{\varphi}^\mu_{je}} & \mathcal{O}_e^{\hat{p}_{\mu-1}} \\ \Big\downarrow{\scriptstyle \hat{\psi}^\mu_{ije}} & & \Big\downarrow{\scriptstyle \hat{\psi}^{\mu-1}_{ije}} \\ \mathcal{O}_e^{\hat{p}_\mu} & \xrightarrow[\hat{\varphi}^\mu_{ie}]{} & \mathcal{O}_e^{\hat{p}_{\mu-1}} \end{array}$$

such that $\hat{\varphi}_{ie}$ resp. $\hat{\psi}^\mu_{ije}$ are defined on $R_i \times S_e$ resp. $R_{ij} \times S_e$ and $\hat{\psi}^\mu_{ije}$ are invertible such that with some constant L_4 resulting by algebraic operations from L_1, L_2, L_3 we have

$$\begin{aligned} &\|\hat{\varphi}^\mu_{ie} - \hat{\varphi}^\mu_{ie}\|_{R_i,\rho} \leq L_4 \varepsilon |\rho|^{e_o} \\ &\|\hat{\psi}^\mu_{ije} - \hat{\psi}^\mu_{ije_o}\|_{R_{ij},\rho} \leq L_4 \varepsilon |\rho|^{e_o} . \end{aligned} \tag{5}$$

Let now $\varepsilon_o(i,j)$ be the ε in Cartan's Lemma (6.11). Then $L_4 \varepsilon |\rho|^{e_o} \leq \varepsilon(i,j)$ only yields a new triangular condition, which we can assume to be satisfied. Now, since $\hat{\psi}^\mu_{ije}$ are invertible, by (6.12) we find invertible matrices Φ^μ_{ie} on $R_i \times S_e$, such that with a new constant L_5 we have

$$\hat{\psi}^\mu_{ije} = \Phi^\mu_{ie} (\Phi^\mu_{je})^{-1}$$

$$\|\Phi^{\mu}_{ie} - \Phi^{\mu}_{ie_o}\|_{R_i\rho} \leq L_5 \varepsilon |\rho|^{e_o} ,$$

if the above estimates are satisfied. Here we can assume that $\Phi^{\mu}_{ie_o}$ is already constructed in the same manner for the e_o-situation.

By applying amalgamation II, (4.3), we obtain now a global resolution on $R_i \cup R_j$ and relating homomorphisms on R_i and R_j with estimates of the above type, since only algebraic operations are involved.

To construct the relating homomorphisms for rectangles of the type $R_i \cup R_j$ and $R_k \cup R_\ell$ we can use amalgamation I, cf. (4.1). To do this with bounds one has to use again (II) (in the case where the resolution exists) with respect to $(R_i \cup R_j) \cap (R_k \cup R_\ell)$ and to shrink a little.

By repeating this process a finite number of times we only have to consider a finite number of prescribed intermediate shrinkings and finite intersections and compositions of triangular conditions. Hence for a given rectangle $R' \subset\subset R$ we obtain a triangular condition T such that for $L_o > 0$ and $(L_o, L, \varepsilon, \Theta) \in T$ the condition of c) is verified.
This completes the proof of (7.4).

(7.5) Main Lemma with partially privileged boundary.

Let (G, Y, j, Ω, D) be a special chart with $G = j^{-1}D$ and let $U = j^{-1}\tilde{U}$, $U' = j^{-1}\tilde{U}'$, $U_i' = j^{-1}\tilde{U}_i'$, $U_i = j^{-1}\tilde{U}_i$ where $\tilde{U}_i' \subset\subset \tilde{U}_i \subset\subset \Omega$ and $\tilde{U}' \subset\subset \tilde{U} \subset\subset \Omega$ are Stein open subsets such that $\mathfrak{U}' = (U_i')$ and $\mathfrak{U} = (U_i)$ cover $\bar{U}$. Let

$$\varphi_{e_o}: \mathcal{O}^q_{e_o} \to \mathcal{O}^p_{e_o}$$

be a homomorphism defined in a neighborhood of $\bar{U}$ such that φ_{e_o} is bounded and reduced and assume that D is a polydomain which is privileged with respect to $\mathcal{O}_X$ and Coker φ_o. Then the following two statements hold.

(I) If $g_{e_o} \in \Gamma(G\cap U\times S_{e_o}, \mathcal{O}^q_{e_o})$, $f_{e_o} = \varphi_{e_o} g_{e_o}$, and if g_{e_o} and f_{e_o} are bounded and reduced, then there exists a triangular condition T depending only on the e_o-situation, such that for any $L_o>0$ and any (L,ε,θ) with $(L_o,L,\varepsilon,\theta) \in T$ the following is true:

If φ_e resp. $f_e \in \Gamma(G\cap U\times S_e, \operatorname{Im} \varphi_e)$ extend φ_{e_o} resp. f_{e_o} on $U\times S_e$ resp. $G\cap U\times S_e$, $e\geq e_o$, such that both are bounded and reduced, Coker φ_e is S_e-flat and such that

$$\|\varphi_e - \varphi_{e_o}\|_{U,\rho} \leq L_o\varepsilon|\rho|^{e_o}$$

$$\|f_e - f_{e_o}\|_{G\cap U,\rho} \leq L_o\varepsilon|\rho|^{e_o}$$

then there exists an extension g_e of $g_{e_o}|G\cap U'\times S_{e_o}$ such that $\varphi_e g_e = f_e$ on $G\cap U'\times S_e$ and

$$\|g_e - g_{e_o}\|_{G\cap U',\rho} \leq L\varepsilon|\rho|^{e_o}.$$

(II) Let $s>0$ and $\eta_{e_o} \in C^{s-1}(G\cap U\cap\mathfrak{U}\times S_{e_o}, \operatorname{Im}\varphi_{e_o})$ and $\xi_{e_o} = \delta\eta_{e_o}$ be given, such that ξ_{e_o} and η_{e_o} are bounded and reduced as power series with respect to the norm $\| \ \|_{G\cap U\cap\mathfrak{U},\rho}$, cf. (6.9). Then there is a triangular condition T depending only on this

<u>e_o-situation such that for any</u> $L_o > 0$ <u>and</u> $(L_o, L, \varepsilon, \theta) \in T$ <u>the following is satisfied.</u>

If φ_e extends φ_{e_o} on $U \times S_{e_o}$ such that φ_e is bounded and reduced and Coker φ_e is S_e-flat and if $\xi_e \in Z^1(G \cap U \cap \mathcal{U} \times S_e, \mathrm{Im}\, \varphi_e)$ extends ξ_{e_o} and is bounded and reduced, and if

$$\|\varphi_e - \varphi_{e_o}\|_{U,\rho} \leq L_o \varepsilon |\rho|^{e_o}$$

$$\|\xi_e - \xi_{e_o}\|_{G \cap U \cap \mathcal{U},\rho} \leq L_o \varepsilon |\rho|$$

then there exists an extension

$\eta_e \in C^{s-1}(G \cap U' \cap \mathcal{U}' \times S_e, \mathrm{Im}\, \varphi_e)$ of η_{e_o} on $G \cap U' \times \mathcal{U}'$ such that $\delta \eta_e = \xi_e | G \cap U' \cap \mathcal{U}'$ and

$$\|\eta_e - \eta_{e_o}\|_{G \cap U' \cap \mathcal{U}',\rho} \leq L \varepsilon |\rho|^{e_o} .$$

Remarks: 1) Again this Lemma is independent of $e \gg e_o$.

2) It is possible that $G \subset\subset U$. In this case we only need norms for φ_e over G and not on U.

Proof: The proof is analogous to that of (7.4), where we use (6.7) in order not to shrink G. Since we are allowed to shrink U, we can apply (7.4), (I) to get a global resolution of φ_e on $\tilde{U}'$ with bounds as in (7.4) c). The rest of the proof (7.4) d), e) can be carried through without shrinking G by using (6.7).

§ 8 Extending homomorphisms with bounds.

As in § 7 we assume that a sequence of ideals $\mathcal{J}_e \subset \mathbb{C}_e\langle t\rangle$, $e \leq e_1$, is given and we assume the same conditions with the same notations to be satisfied. We are going to extend global homomorphisms from S_e to S_{e+1}, which are given as in the process of amalgamation II, (4.3). However the following extension process is done without using Cartan's Lemma by simply constructing the coefficients of order e+1 of the reduced power series. We assume the following to be given:

(i) a coherent S_{e+1}-flat sheaf $\mathcal{F}_{e+1}$ on $X \times S_{e+1}$, $e+1 \leq e_1$. We will denote its restriction to $X \times S_d$, $d \leq e+1$, by $\mathcal{F}_d = \mathcal{F}_{e+1} | S_d$ and its restriction to X by $\mathcal{F}_o$.

(ii) a special chart $\mathfrak{U} = (G, Y, j, \Omega, D)$ such that $G = j^{-1}D$, $D \subset\subset \Omega$ and D is a polydomain which is privileged for $\mathcal{O}_X$ and $\mathcal{F}_o$.

(iii) finite Stein coverings $(\tilde{U}_i')$ and $(\tilde{U}_i)$ of $\bar{D}$ such that $\tilde{U}_i' \subset\subset \tilde{U}_i \subset\subset \Omega$. By $\mathcal{U}' = (U_i')$, $\mathcal{U} = (U_i)$ we denote the coverings of $\bar{G}$ given by $U_i' = j^{-1}\tilde{U}_i'$, $U_i = j^{-1}\tilde{U}_i$.

(iv) homomorphisms σ_{ie_o} on $U_i^* \times S_{e_o}$, φ_{ije_o} on $U_{ij}^* \times S_{e_o}$, σ_{e_o} on $G^* \times S_{e_o}$ and $\varphi_{ie_o}, \check{\varphi}_{ie_o}$ on $G^* \cap U_i^* \times S_{e_o}$, where $G \subset\subset G^*$ and $U_i \subset\subset U_i^*$, such that the following diagram is commutative with exact rows

$$
\begin{array}{ccccccc}
\mathcal{O}^{q}_{e_o} & \xrightarrow{\sigma_{e_o}} & \mathcal{O}^{p}_{e_o} & \longrightarrow & \mathcal{F}_{e_o} & \longrightarrow & 0 \\
{\scriptstyle \psi_{ie_o}} \Big\downarrow & & {\scriptstyle \check{\varphi}_{ie_o}} \Big\uparrow \Big\downarrow {\scriptstyle \varphi_{ie_o}} & & \Big\| & & \\
\mathcal{O}^{q_i}_{e_o} & \xrightarrow{\sigma_{ie_o}} & \mathcal{O}^{p_i}_{e_o} & \longrightarrow & \mathcal{F}_{e_o} & \longrightarrow & 0 \\
 & & \Big\downarrow {\scriptstyle \varphi_{jie_o}} & & \Big\| & & \\
\mathcal{O}^{q_j}_{e_o} & \xrightarrow{\sigma_{je_o}} & \mathcal{O}^{p_j}_{e_o} & \longrightarrow & \mathcal{F}_{e_o} & \longrightarrow & 0
\end{array}
$$

Moreover we assume given ψ_{ie_o}, g_{ie_o}, f_{jie_o} with respect to $G^* \cap \mathfrak{U}^*$ such that we have the identities

$$\varphi_{ie_o}\sigma_{e_o} = \sigma_{ie_o}\psi_{ie_o}$$

$$\varphi_{jie_o}\varphi_{ie_o} = \varphi_{je_o} + \sigma_{je_o}f_{jie_o}$$

$$\varphi_{ie_o}\check{\varphi}_{ie_o} = \mathrm{id} + \sigma_{ie_o}g_{ie_o} \quad .$$

We assume that all these homomorphisms are reduced and bounded with respect to $\mathfrak{U}$ resp. $G \cap \mathfrak{U}$. Under these assumptions we obtain

<u>(8.1) Lemma:</u> <u>For any pair of constants</u> L_o, $L>0$ <u>there is a constant</u> L_o' <u>depending only on</u> L_o <u>and the</u> 0-<u>situation and there is a triangular condition</u> E <u>for</u> ε, θ, <u>depending on</u> L <u>and the</u> e_o-<u>situation, such that for</u> $(\varepsilon,\theta) \in E$ <u>the following holds.</u>

Assume that $\sigma_{i,e+1}$, $\varphi_{ji,e+1}$ *extend* σ_{ie_o}, φ_{ji,e_o} *over* $U_i\times S_{e+1}$ *and* $U_{ij}\times S_{e+1}$ *and that* σ_e, φ_{ie}, $\check{\varphi}_{ie}$ *extend* σ_{e_o}, φ_{ie_o}, $\check{\varphi}_{ie_o}$ *over* $G\times S_e$, $G\cap U_i\times S_e$ *such that we obtain the commutative diagrams with exact rows*

$$\begin{array}{ccccccc}
\mathcal{O}_e^q & \xrightarrow{\sigma_e} & \mathcal{O}_e^p & \longrightarrow & \mathcal{F}_e & \longrightarrow & 0\\
 & & \check{\varphi}_{ie}\Big\uparrow\Big\downarrow\varphi_{ie} & & \Big\| & & \\
\mathcal{O}_e^{q_i} & \xrightarrow{\sigma_{ie}} & \mathcal{O}_e^{p_i} & \longrightarrow & \mathcal{F}_e & \longrightarrow & 0
\end{array}$$

$$\begin{array}{ccccccc}
\mathcal{O}_{e+1}^{q_i} & \xrightarrow{\sigma_{i,e+1}} & \mathcal{O}_{e+1}^{p_i} & \longrightarrow & \mathcal{F}_{e+1} & \longrightarrow & 0\\
 & & \Big\downarrow\varphi_{ji,e+1} & & \Big\| & & \\
\mathcal{O}_{e+1}^{q_j} & \xrightarrow{\sigma_{j,e+1}} & \mathcal{O}_{e+1}^{p_j} & \longrightarrow & \mathcal{F}_{e+1} & \longrightarrow & 0\ ,
\end{array}$$

where $\sigma_{ie}=\sigma_{i,e+1}|S_e$ *and* $\varphi_{jie}=\varphi_{ji,e+1}|S_e$ *and such that all these homomorphisms are bounded and reduced with respect to* $G\cap\mathfrak{U}$ *and satisfy the estimates*

$$\left.\begin{array}{l}\|\sigma_{i,e+1}-\sigma_{ie}\|_{U_i,\rho}\\[2ex] \|\varphi_{ji,e+1}-\varphi_{jie}\|_{U_{ij},\rho}\end{array}\right\}\le L_o\varepsilon|\rho|^{e_o}\qquad \left.\begin{array}{l}\|\sigma_{ie}-\sigma_{ie_o}\|_{U_i\rho}\\[2ex] \|\varphi_{jie}-\varphi_{jie_o}\|_{U_{ij}\rho}\end{array}\right\}\le L\varepsilon|\rho|^{e_o}$$

$$\left.\begin{array}{l} \|\sigma_e - \sigma_{e_o}\|_{G,\rho} \\[2ex] \|\varphi_{ie} - \varphi_{ie_o}\|_{G\cap U_i,\rho} \\[2ex] \|\check{\varphi}_{ie} - \check{\varphi}_{ie_o}\|_{G\cap U_i,\rho} \end{array}\right\} \leq L\varepsilon|\rho|^{e_o}$$

Then there exist extensions $\sigma_{e+1}, \varphi_{i,e+1}, \check{\varphi}_{i,e+1}$ of $\sigma_e, \varphi_{ie}, \check{\varphi}_{ie}$ *respectively, defined on* G resp. $G\cap U_i'$, *such that we have a commutative diagram with exact rows*

$$\begin{array}{ccccccc} \mathcal{O}^{q}_{e+1} & \xrightarrow[\sigma_{e+1}]{} & \mathcal{O}^{p}_{e+1} & \longrightarrow & \mathcal{F}_{e+1} & \longrightarrow & 0 \\ & & {\scriptstyle \check{\varphi}_{i,e+1}}\uparrow\Big\| \downarrow{\scriptstyle \varphi_{i,e+1}} & & \Big\| & & \\ \mathcal{O}^{q_i}_{e+1} & \xrightarrow[\sigma_{i,e+1}]{} & \mathcal{O}^{p_i}_{e+1} & \longrightarrow & \mathcal{F}_{e+1} & \longrightarrow & 0\,, \end{array}$$

and such that the following estimates hold:

$$\left.\begin{array}{l} \|\sigma_{e+1} - \sigma_e\|_{G,\rho} \\[2ex] \|\varphi_{i,e+1} - \varphi_{ie}\|_{G\cap U_i',\rho} \\[2ex] \|\check{\varphi}_{i,e+1} - \check{\varphi}_{ie}\|_{G\cap U_i',\rho} \end{array}\right\} \leq L_o'\varepsilon|\rho|^{e_o} \quad .$$

Remarks: 1) Only $\mathfrak{U}$ is shrunken but not G.

2) The constants and estimates are independent of $e \geq e_o$.

3) There is a version of (8.1) with extensions constructed directly from S_{e_o} to S_e as in § 7.

4) If G is not privileged one can prove an analogue Lemma, where G and $\mathfrak{U}$ are shrunk.

Proof: a) Restricting all given homomorphisms to X one obtains the induced diagram in which we denote by π resp. π_i the corresponding epimorphisms onto $\mathcal{F}_o$.

$$\begin{array}{ccccccc}
\mathcal{O}_X^q & \xrightarrow[\sigma_o]{} & \mathcal{O}_X^p & \xrightarrow[\pi]{} & \mathcal{F}_o & \longrightarrow & 0 \\
 & & \Big\downarrow{\scriptstyle \varphi_{io}} & & \Big\| & & \\
\mathcal{O}_X^{q_i} & \xrightarrow[\sigma_{i_o}]{} & \mathcal{O}_X^{p_i} & \xrightarrow[\pi_i]{} & \mathcal{F}_o & \longrightarrow & 0 \\
 & & \Big\downarrow{\scriptstyle \varphi_{jio}} & & \Big\| & & \\
\mathcal{O}_X^{q_j} & \xrightarrow[\sigma_{jo}]{} & \mathcal{O}_X^{p_j} & \xrightarrow[\pi_j]{} & \mathcal{F}_o & \longrightarrow & 0 .
\end{array}$$

First let $U_i' \subset\subset V_i' \subset\subset V_i \subset\subset U_i$ define intermediate coverings and consider pairs of cochains

$$\xi \in C^o(G \cap \mathfrak{W}, \mathcal{H}om(\mathcal{O}_X^q, \mathcal{F}_o))$$
$$\eta \in Z^1(G \cap \mathfrak{W}, \mathcal{H}om(\mathcal{O}_X^p, \mathcal{F}_o))$$

such that $\eta \circ \sigma_o = \delta\xi$.

By (6.8), (ii) there is a constant K_o such that for any such ξ, η we can find ζ in $C^o(G \cap \mathfrak{W}', \mathcal{H}om(\mathcal{O}_X^p, \mathcal{F}_o))$ satisfying $\delta\zeta = \eta | G \cap \mathfrak{W}'$ and $\|\zeta\|^{\pi}_{G \cap \mathfrak{W}'} \leq K_o \|\eta\|^{\pi}_{G \cap \mathfrak{W}}$. Then $s = \xi - \zeta\sigma_o$ defines a section in $\Gamma(G, \mathcal{H}om(\mathcal{O}_X^q, \mathcal{F}_o))$ such that

$$\|s\|_G \leq K_o \|\xi\|_{G \cap \mathfrak{W}} + K_o^2 M_o \|\eta\|_{G \cap \mathfrak{W}} \quad .$$

Here M_o denotes the maximum of the norms of all homomorphisms $\sigma_{e_o}, \ldots, g_{ie_o}$ of the e_o-situation and $\sigma_o, \ldots, g_{io}$ of the O-situation.

b) Let now $L_o > 0$ be arbitrary and let L_o' satisfy

$$\left.\begin{array}{l} K_o^3(3 + M_oL_o) \\ 2K_o(3 + M_oL_o + 3M_oK_o + 2M_o^2K_oL_o) \\ 2K_o(3 + M_oL_o) + K_o^3M_o(3 + 2L_oM_o) \end{array}\right\} \leq L_o'$$

and put $L_1 = 2L^2 + 4M_oL + 1$. By (7.5), (I) there are triangular conditions T_i for the homomorphisms σ_{ie_o} and the geometric situation G, $V_i \subset\subset U_i$ with e_o-data. Let L^* be defined as the minimal value of all the L_1^* satisfying $(L_1, L_1^*, \varepsilon, \Theta) \in \cap T_i$. Then L^* only depends on the choice of L. Now restrict $\varepsilon > 0$ to the condition

$$2\varepsilon(3L + L^* + L^2 + LL^*) \leq 1$$

$$2\varepsilon(2L_1 + L_1^2) \leq 1$$

and after that we choose Θ so small that the following estimates are verified:

$$\|\sigma_{ie_o} - \sigma_{io}\|_{U_i,\rho} \leq \varepsilon \qquad \|\sigma_{e_o} - \sigma_o\|_{G,\rho} \leq \varepsilon$$

$$\|\varphi_{ji,e_o} - \varphi_{jio}\|_{U_{ij},\rho} \leq \varepsilon \qquad \|\varphi_{ie_o} - \varphi_{io}\|_{G\cap U_i,\rho} \leq \varepsilon$$

$$\|\check{\varphi}_{ie_o} - \check{\varphi}_{io}\|_{G\cap U_i,\rho} \leq \varepsilon$$

$$\| \psi_{ie_o} - \psi_{io} \|_{G\cap U_i,\rho} \leq \varepsilon$$

$$\| g_{ie_o} - g_{io} \|_{G\cap U_i,\rho} \leq \varepsilon$$

$$\| f_{jie_o} - f_{jio} \|_{G\cap U_{ij},\rho} \leq \varepsilon$$

and independent of $e \geq e_o$, cf. (5.7.2) and (5.1):

$$\| \mathrm{red}_e(\sigma_{ie_o}\psi_{ie_o} - \varphi_{ie_o}\sigma_{e_o}) \|_{G\cap U_i,\rho} \leq \varepsilon|\rho|^{e_o}$$

$$\| \mathrm{red}_e(\sigma_{je_o}f_{jie_o} + \varphi_{je_o} - \varphi_{jie_o}\varphi_{ie_o}) \|_{G\cap U_{ij},\rho} \leq \varepsilon|\rho|^{e_o}$$

$$\| \mathrm{red}_e(\sigma_{ie_o}g_{ie_o} + \mathrm{id} - \varphi_{ie_o}\varphi_{ie_o}) \|_{G\cap U_i,\rho} \leq \varepsilon|\rho|^{e_o}.$$

All these conditions yield a triangular condition, E for ε,θ, which is the condition wanted. Let now $(\varepsilon,\theta) \in E$.

c) Next we construct extensions ψ_{ie}, f_{jie}, g_{ie} with bounds of the auxiliary homomorphism by using (7.5) (I) and the triangular condition T_i. For this reason we first estimate $\varphi_{ie}\sigma_e$, $\varphi_{jie}\varphi_{ie}$, $\varphi_{ie}\check{\varphi}_{ie}$. We can assume that the triangular condition E also yields the estimates

$$\| \mathrm{red}_e\varphi_{ie_o}\sigma_{e_o} - \mathrm{red}_{e_o}\varphi_{ie_o}\sigma_{e_o} \|_{G\cap U_i,\rho} \leq \varepsilon|\rho|^{e_o}$$

$$\| \mathrm{red}_e\varphi_{jie_o}\varphi_{ie_o} - \mathrm{red}_{e_o}\varphi_{jie_o}\varphi_{ie_o} \|_{G\cap U_{ij},\rho} \leq \varepsilon|\rho|^{e_o}$$

$$\| \mathrm{red}_e\varphi_{ie_o}\check{\varphi}_{ie_o} - \mathrm{red}_{e_o}\varphi_{ie_o}\check{\varphi}_{ie_o} \|_{G\cap U_i,\rho} \leq \varepsilon|\rho|^{e_o}.$$

Then we obtain

$$\| \mathrm{red}_e\varphi_{ie}\sigma_e - \mathrm{red}_{e_o}\varphi_{ie_o}\sigma_{e_o} \|_{G\cap U_i,\rho}$$

$$\leq \|red_e(\varphi_{ie}\sigma_e - \varphi_{ie_o}\sigma_{e_o})\|_{G\cap U_i,\rho} + \varepsilon|\rho|^{e_o}$$

$$\leq 2[L^2\varepsilon^2|\rho|^{2e_o} + 2M_oL\varepsilon|\rho|^{e_o}] + \varepsilon|\rho|^{e_o}$$

where we used the product decomposition into

$(\varphi_{ie} - \varphi_{ie_o})(\sigma_e - \sigma_{e_o}) + (\varphi_{ie} - \varphi_{ie_o})\sigma_{e_o} + \varphi_{ie_o}(\sigma_e - \sigma_{e_o})$

and the estimate in the assumption of the Lemma. Since we can also assume $\varepsilon|\rho|^{e_o} \leq 1$, by the definition of L_1 we obtain

$$\|red_e\varphi_{ie}\sigma_e - red_{e_o}\varphi_{ie_o}\sigma_{e_o}\|_{G\cap U_i,\rho} \leq L_1\varepsilon|\rho|^{e_o} .$$

Similarly

$$\|red_e\varphi_{jie}\varphi_{ie} - red_{e_o}\varphi_{jie_o}\varphi_{ie_o}\|_{G\cap U_{ij},\rho} \leq L_1\varepsilon|\rho|^{e_o}$$

$$\|red_e\varphi_{ie}\check{\varphi}_{ie} - red_{e_o}\varphi_{ie_o}\check{\varphi}_{ie_o}\|_{G\cap U_i,\rho} \leq L_1\varepsilon|\rho|^{e_o} .$$

Now by $(L_1,L^*,\varepsilon,\Theta) \in \cap T_i$, cf. b), by (7.5), (I) we can find ψ_{ie}, f_{jie}, g_{ie} extending ψ_{ie_o}, f_{jie_o}, g_{ie_o} over $G\cap V_i$ and $G\cap V_{ij}$, satisfying

$$\sigma_{ie}\psi_{ie} = \varphi_{ie}\sigma_e \qquad \text{on } G\cap V_i\times S_e \qquad (1a)$$

$$\sigma_{je}f_{jie} = \varphi_{jie}\varphi_{ie} - \varphi_{je} \qquad \text{on } G\cap V_{ij}\times S_e \qquad (1b)$$

$$\sigma_{ie}g_{ie} = \varphi_{ie}\check{\varphi}_{ie} - id \qquad \text{on } G\cap V_i\times S_e \qquad (1c)$$

and such that

$$\|\psi_{ie} - \psi_{ie_o}\|_{G\cap V_i,\rho} \leq L^*\varepsilon|\rho|^{e_o}$$

$$\|f_{jie} - f_{jie_o}\|_{G\cap V_{ij},\rho} \leq L^*\varepsilon|\rho|^{e_o}$$

$$\|g_{ie} - g_{ie_o}\|_{G\cap V_i,\rho} \leq L^*\varepsilon|\rho|^{e_o} .$$

d) After this construction we obtain the estimates

$$\|\mathrm{red}_{e+1}(\sigma_{ie}\psi_{ie} - \varphi_{ie}\sigma_e)\|_{G\cap V_i,\rho} \leq 3\varepsilon|\rho|^{e_o} \tag{2a}$$

$$\|\mathrm{red}_{e+1}(\sigma_{je}f_{jie} + \varphi_{je} - \varphi_{jie}\varphi_{ie})\|_{G\cap V_{ij},\rho} \leq 3\varepsilon|\rho|^{e_o} \tag{2b}$$

$$\|\mathrm{red}_{e+1}(\sigma_{ie}g_{ie} + \mathrm{id} - \varphi_{ie}\check{\varphi}_{ie})\|_{G\cap V_i,\rho} \leq 3\varepsilon|\rho|^{e_o} . \tag{2c}$$

These estimates are obtained by writing for example

$$\mathrm{red}_{e+1}(\sigma_{ie}\psi_{ie} - \varphi_{ie}\sigma_e) = \mathrm{red}^{e+1}(\sigma_{ie}\psi_{ie} - \varphi_{ie}\sigma_e)$$

$$= \mathrm{red}^{e+1}(\sigma_{ie}\psi_{ie} - \sigma_{ie_o}\psi_{ie_o}) + \mathrm{red}^{e+1}(\varphi_{ie_o}\sigma_{e_o} - \varphi_{ie}\sigma_e)$$

$$+ \mathrm{red}^{e+1}(\sigma_{ie_o}\psi_{ie_o} - \varphi_{ie_o}\sigma_{e_o})$$

and using (5.8), by which we can estimate the first two terms, where as the third is already $\leq \varepsilon|\rho|^{e_o}$. Since the relations (1a), (1b), (1c) are satisfied the terms in (2a), (2b), (2c) only have order e+1. Thus for (2b) resp. (2c) we can neglect φ_{je} resp. id. We define the coefficients $a_{i\nu}$, $b_{ji\nu}$, $c_{i\nu}$ by

$$\sum_\nu a_{i\nu}t^\nu = \mathrm{red}_{e+1}(\sigma_{ie}\psi_{ie} - \varphi_{ie}\sigma_e) \tag{3a}$$

$$\sum_\nu b_{ji\nu}t^\nu = \mathrm{red}_{e+1}(\sigma_{je}f_{jie} + \varphi_{je} - \varphi_{jie}\varphi_{ie}) \tag{3b}$$

$$\sum_\nu c_{i\nu}t^\nu = \mathrm{red}_{e+1}(\sigma_{ie}g_{ie} + \mathrm{id} - \varphi_{ie}\varphi_{ie}) \tag{3c}$$

where summation is over all $\nu \in \Delta_{e+1}$.

e) Let us now assume that extensions $\sigma_{i,e+1}$, $\varphi_{ij,e+1}$ are given as in the Lemma. If $\sigma_{i\nu}$, $\varphi_{ji\nu}$ are the coefficients of order e+1 of these extensions, $\nu \in \Delta_{e+1}$, we have $\varphi^\nu\|\sigma_{i\nu}\|_{U_i} \leq \gamma_\nu L_o\varepsilon|\rho|^{e_o}$ and $\rho^\nu\|\varphi_{ji\nu}\|_{U_{ij}} \leq \gamma_\nu L_o\varepsilon|\rho|^{e_o}$. We have to construct coefficients

σ_ν, $\varphi_{i\nu}$, $\check{\varphi}_{i\nu}$ and $\psi_{i\nu}$, $f_{ji\nu}$, $g_{i\nu}$ on G, $G \cap U_i^!$, $G \cap U_{ij}^!$ respectively such that the following equations are satisfied mod $\widetilde{\mathfrak{J}}_{e+1}$.

$$(\sigma_{ie} + \Sigma\sigma_{i\nu}t^\nu)(\psi_{ie}+\Sigma\psi_{i\nu}t^\nu) \equiv (\varphi_{ie}+\Sigma\varphi_{i\nu}t^\nu)(\sigma_e+\Sigma\sigma_\nu t^\nu) \tag{4a}$$

$$(\varphi_{jie}+\Sigma\varphi_{ji\nu}t^\nu)(\varphi_{ie}+\Sigma\varphi_{i\nu}t^\nu) \equiv \varphi_{je}+\Sigma\varphi_{j\nu}t^\nu+(\sigma_{je}+\Sigma\sigma_{j\nu}t^\nu)(f_{jie}+\Sigma f_{ji\nu}t^\nu) \tag{4b}$$

$$(\varphi_{ie}+\Sigma\varphi_{i\nu}t^\nu)(\check{\varphi}_{ie}+\Sigma\check{\varphi}_{i\nu}t^\nu) \equiv 1+(\sigma_{ie}+\Sigma\sigma_{i\nu}t^\nu)(g_{ie}+\Sigma g_{i\nu}t^\nu) \tag{4c}$$

where again summation is over $\nu \in \Delta_{e+1}$. Since $\mathfrak{m}^{e+2} \subset \widetilde{\mathfrak{J}}_{e+1}$ and since the coefficients are unique these equations are equivalent to

$$\sigma_{io}\psi_{i\nu} + \sigma_{i\nu}\psi_{io} + a_{i\nu} = \varphi_{io}\sigma_\nu + \varphi_{i\nu}\sigma_o , \tag{5a}$$

$$\varphi_{jio}\varphi_{i\nu} + \varphi_{ji\nu}\varphi_{io} = \varphi_{j\nu} + b_{ji\nu} + \sigma_{jo}f_{ji\nu} + \sigma_{j\nu}f_{jio} , \tag{5b}$$

$$\varphi_{io}\check{\varphi}_{i\nu} + \varphi_{i\nu}\check{\varphi}_{io} = c_{i\nu} + \sigma_{io}g_{i\nu} + \sigma_{i\nu}g_{io} . \tag{5c}$$

Applying π_i or π_j to both sides we obtain

$$\xi_{i\nu} := \pi_i a_{i\nu} + \pi_i\sigma_{i\nu}\psi_{io} = \pi\sigma_\nu + \pi_i\varphi_{i\nu}\sigma_o \tag{6a}$$

$$\eta_{ji\nu} := \pi_j b_{ji\nu} + \pi_j\sigma_{j\nu}f_{jio} - \pi_j\varphi_{ji\nu}\varphi_{io} = \pi_i\varphi_{i\nu} - \pi_j\varphi_{j\nu} \tag{6b}$$

$$z_{i\nu} := \pi_i c_{i\nu} + \pi_i\sigma_{i\nu}g_{io} = \pi_i\varphi_{i\nu}\check{\varphi}_{io} + \pi_i\varphi_{i\nu} \tag{6c}$$

f) The left hand sides are defined at least on $G \cap V_i$ resp. $G \cap V_{ij}$. Let $\xi_\nu = (\xi_{i\nu})$ and $\eta_\nu = (\eta_{ji\nu})$ be the Čech-cochains with respect to $G \cap \mathfrak{W}$, $\nu \in \Delta_{e+1}$. Since on $G \cap \mathfrak{U}$ we can find algebraic extensions of σ_e, φ_{ie}, $\check{\varphi}_{ie}$ and ψ_{ie}, f_{jie}, g_{ie} and since on $G' \cap \mathfrak{U}'$ these extension can be reduced with bounded coefficients, we do have relations (6) for any $G' \subset\subset G$ and any $\mathfrak{U}' \subset\subset \mathfrak{W}$. This implies $\delta\eta_\nu = 0$, $\delta\xi_\nu = \eta_\nu \circ \sigma_o$ on $G \cap \mathfrak{W}$. Now by a) we can find

$\zeta_\nu \in C^0(G\cap\mathcal{W}', \mathcal{H}om(\mathcal{O}_X^p, \mathcal{F}_0))$ and a section $s_\nu \in \Gamma(G, \mathcal{H}om(\mathcal{O}_X^q, \mathcal{F}_0))$ such that $s_\nu = \xi_\nu - \zeta_\nu\sigma_0$ and $\eta_\nu = \delta\zeta_\nu$ and

$$\|\zeta_\nu\|^\pi_{G\cap\mathcal{W}'} \leq K_0\|\eta_\nu\|^\pi_{G\cap\mathcal{W}}$$

$$\|s_\nu\|^\pi_G \leq K_0\|\xi_\nu\|^\pi_{G\cap\mathcal{W}} + K_0^2 M_0\|\eta_\nu\|^\pi_{G\cap\mathcal{W}} .$$

Since φ_{io}, π_i, π are defined with respect to the larger covering $G^*\cap\mathcal{U}^*$, φ_{io} is bounded on $G\cap U_i'$ and hence $\zeta_{i\nu}$ can be estimated with respect to the quotient norm of π_i, cf. (6.8), for which we also use the constant K_0. Hence there are $\varphi_{i\nu} \in \Gamma(G\cap U_i', \mathcal{H}om(\mathcal{O}_X^p, \mathcal{O}_X^{p_i}))$ such that $\pi_i\varphi_{i\nu} = \zeta_{i\nu}$ and

$\|\varphi_{i\nu}\|_{G\cap U_i'} \leq K_0\|\zeta_\nu\|^\pi_{G\cap\mathcal{W}'}$. Also we find some $\sigma_\nu \in \Gamma(G, \mathcal{H}om(\mathcal{O}_X^q, \mathcal{O}_X^p))$ such that $\pi\sigma_\nu = s_\nu$ and $\|\sigma_\nu\|_G \leq 2\|s_\nu\|^\pi_G$. Moreover there are also $\check{\varphi}_{i\nu}$ satisfying $z_{i\nu} - \pi_i\varphi_{i\nu}\check{\varphi}_{io} = \pi_i\check{\varphi}_{i\nu}$ and $\|\check{\varphi}_{i\nu}\|_{G\cap U_i'} \leq 2\|z_{i\nu}\|_{G\cap V_i} + M_0\|\varphi_{i\nu}\|_{G\cap U_i'}$.

With the now defined σ_ν, $\varphi_{i\nu}$, $\check{\varphi}_{i\nu}$ the relations (6) are satisfied with respect to $G\cap\mathcal{U}'$. By Theorem B we can find $\psi_{i\nu}$, $f_{ji\nu}$, $g_{i\nu}$ not necessarily with bounds such that (5) is satisfied. Defining finally

$$\sigma_{e+1} = \sigma_e + \Sigma\,\sigma_\nu t^\nu \qquad \psi_{i,e+1} = \psi_{ie} + \Sigma\,\psi_{i\nu}t^\nu$$

$$\varphi_{i,e+1} = \varphi_{ie} + \Sigma\,\varphi_{i\nu}t^\nu \qquad f_{ji,e+1} = f_{jie} + \Sigma\,f_{ji\nu}t^\nu$$

$$\check{\varphi}_{i,e+1} = \check{\varphi}_{ie} + \Sigma\,\check{\varphi}_{i\nu}t^\nu \qquad g_{i,e+1} = g_{ie} + \Sigma\,g_{i\nu}t^\nu$$

we obtain the desired extensions with

$$\sigma_{i,e+1}\psi_{i,e+1} = \varphi_{i,e+1}\sigma_{e+1} \tag{7a}$$

$$\varphi_{ji,e+1}\varphi_{i,e+1} = \varphi_{j,e+1} + \sigma_{j,e+1}f_{ji,e+1} \tag{7b}$$

$$\varphi_{i,e+1}\check{\varphi}_{i,e+1} = \mathrm{id} + \sigma_{i,e+1}g_{i,e+1} \ . \tag{7c}$$

g) To get the final estimates of $\sigma_{e+1} - \sigma_{e_o}$ and $\varphi_{i,e+1} - \varphi_{ie_o}$ we estimate $\xi_{i\nu}$, $\eta_{ji\nu}$, $z_{i\nu}$ by their definitions (6) and use the conditions on the constants. By the assumed bounds for $\sigma_{i\nu}$ and φ_{ji}, cf. e), and by (2), (3) we have

$$\rho^\nu \|\xi_{i\nu}\|^{\pi_i}_{G\cap V_i} \leq \gamma_\nu \varepsilon |\rho|^{e_o}(3 + L_oM_o)$$

$$\rho^\nu \|\eta_{ji\nu}\|^{\pi_j}_{G\cap V_{ij}} \leq \gamma_\nu \varepsilon |\rho|^{e_o}(3 + 2L_oM_o)$$

$$\rho^\nu \|z_{i\nu}\|^{\pi_i}_{G\cap V_i} \leq \gamma_\nu \varepsilon |\rho|^{e_o}(3 + L_oM_o) \ .$$

Taking the quotient norms with respect to π we have to multiply with K_o, cf. (9.8), such that

$$\rho^\nu \|\xi_\nu\|^{\pi}_{G\cap\mathfrak{V}} \leq \gamma_\nu \varepsilon |\rho|^{e_o}K_o(3 + L_oM_o)$$

$$\rho^\nu \|\eta_\nu\|^{\pi}_{G\cap\mathfrak{V}} \leq \gamma_\nu \varepsilon |\rho|^{e_o}K_o(3 + 2L_oM_o)$$

$$\rho^\nu \|z_{i\nu}\|^{\pi}_{G\cap V_i} \leq \gamma_\nu \varepsilon |\rho|^{e_o}K_o(3 + L_oM_o) \ .$$

From these estimates we obtain

$$\rho^\nu \|\varphi_{i\nu}\|_{G\cap U_i'} \leq \gamma_\nu \varepsilon |\rho|^{e_o}K_o^3(3 + 2L_oM_o) \leq \gamma_\nu \varepsilon |\rho|^{e_o}L_o'$$

$$\rho^\nu \|\sigma_\nu\|_G \leq \gamma_\nu \varepsilon |\rho|^{e_o}2K_o^2(3 + L_oM_o + K_oM_o(3 + 2L_oM_o)) \leq \gamma_\nu \varepsilon |\rho|^{e_o}L_o'$$

$$\rho^\nu \|\check{\varphi}_{i\nu}\|_{G\cap U_i'} \leq \gamma_\nu \varepsilon |\rho|^{e_o}K_o(2(3 + L_oM_o) + M_oK_o^2(3 + 2L_oM_o)) \leq \gamma_\nu \varepsilon |\rho|^{e_o}L_o'$$

By our choice of the constant L_o' in b) we obtain the desired estimate.

h) Finally we show that we can assume that the homomorphisms induced for the cokernels by $\varphi_{i,e+1}$ and $\check{\varphi}_{i,e+1}$ are the identity. First $\bar{\varphi}_{i,e+1}$: Coker $\sigma_{e+1} \longrightarrow \mathcal{F}_{e+1}$ is an isomorphism since $\bar{\varphi}_{i,e+1}|S_e = \mathrm{id}$ and $\mathcal{F}_{e+1}$ is S_{e+1}-flat. Since $\bar{\varphi}_{ij,e+1} = \mathrm{id}$ by assumption, we do have $\bar{\varphi}_{i,e+1} = \bar{\varphi}_{j,e+1}$ on $G \cap U'_{ij} \times S_{e+1}$ and hence a global isomorphism γ : Coker $\sigma_{e+1} \longrightarrow \mathcal{F}_{e+1}$ is defined on G. If we define the epimorphism $\mathcal{O}^p_{e+1} \longrightarrow \mathcal{F}_{e+1}$ on $G \times S_{e+1}$ by $\mathcal{O}^p_{e+1} \longrightarrow \text{Coker } \sigma_{e+1} \xrightarrow[\gamma]{} \mathcal{F}_{e+1}$, $\varphi_{i,e+1}$ induces the identity for $\mathcal{F}_{e+1}$. The relation (7c) finally shows that $\check{\varphi}_{i,e+1}$ induces a homomorphism for the cokernels, which then must be the identity. This completes the proof of (8.1).

(8.2) In §10 we also need amalgamation I with estimates for (e+1)-order terms. For this reason we prove an analogue of Lemma (8.1) in the situation of amalgamation I. Let the assumptions (i), (ii), (iii) of the beginning of this paragraphe be satisfied. Instead of (iv) we now assume the following to be given:

(iv*) homomorphisms σ_{1e_o}, σ_{2e_o}, φ_{e_o} over $G^* \times S_{e_o}$ and φ_{ie_o} over $G^* \cap U^*_i \times S_{e_o}$, where $G \subset\subset G^*$ and $U_i \subset\subset U^*_i$ such that the following diagram is commutative with exact rows

$$\begin{array}{ccccccc}
\mathcal{O}^q_{e_o} & \xrightarrow{\sigma_{1e_o}} & \mathcal{O}^p_{e_o} & \longrightarrow & \mathcal{F}_{e_o} & \longrightarrow & 0 \\
 & & {\scriptstyle \varphi_{e_o}}\downarrow\downarrow{\scriptstyle \varphi_{ie_o}} & & \| & & \\
\mathcal{O}^q_{e_o} & \xrightarrow{\sigma_{2e_o}} & \mathcal{O}^p_{e_o} & \longrightarrow & \mathcal{F}_{e_o} & \longrightarrow & 0 .
\end{array}$$

We also assume that homomorphisms g_{ie_o} over $G^*\cap U_i^*\times S_{e_o}$ are given such that

$$\varphi_{e_o} - \varphi_{ie_o} = \sigma_{2e_o} g_{ie_o}.$$

We assume that all homomorphisms are bounded and reduced on G resp. $G\cap U_i$.

(8.3) Lemma: For any pair of constants L_o, $L>0$ there is a constant L_o' depending only on L_o and the 0-situation and there is a triangular condition E for ε,θ, depending on L and the e_o-situation, such that for $(\varepsilon,\theta)\in E$ the following holds. If $\sigma_{k,e+1}$, $\varphi_{i,e+1}$, φ_e extend σ_{ke_o}, φ_{ie_o}, φ_{e_o} over G, $G\cap U_i$ respectively such that the diagrams

$$\begin{array}{ccccccc}
\mathcal{O}_e^q & \xrightarrow{\sigma_{1e}} & \mathcal{O}_e^p & \longrightarrow & \mathcal{F}_e & \longrightarrow & 0 \\
 & & \varphi_e \downarrow\downarrow \varphi_{ie} & & \| & & \\
\mathcal{O}_e^q & \xrightarrow{\sigma_{2e}} & \mathcal{O}_e^p & \longrightarrow & \mathcal{F}_e & \longrightarrow & 0
\end{array}$$

$$\begin{array}{ccccccc}
\mathcal{O}_{e+1}^q & \xrightarrow{\sigma_{1,e+1}} & \mathcal{O}_{e+1}^p & \longrightarrow & \mathcal{F}_{e+1} & \longrightarrow & 0 \\
 & & \downarrow \varphi_{i,e+1} & & \| & & \\
\mathcal{O}_{e+1}^q & \xrightarrow{\sigma_{2,e+1}} & \mathcal{O}_{e+1}^p & \longrightarrow & \mathcal{F}_{e+1} & \longrightarrow & 0
\end{array}$$

are commutative with exact rows, where $\sigma_{ke} = \sigma_{k,e+1}|S_e$

$\varphi_{ie}=\varphi_{i,e+1}|S_e$, and if all these homomorphisms are bounded and reduced with respect to G or $G\cap U_i$ and satisfy the estimates

$$\|\sigma_{ke} - \sigma_{ke_o}\|_{G,\rho} \le L\varepsilon|\rho|^{e_o} \qquad \|\sigma_{k,e+1} - \sigma_{ke}\|_{G,\rho} \le L_o\varepsilon|\rho|^{e_o}$$

$$\|\varphi_{ie} - \varphi_{ie_o}\|_{G\cap U_i,\rho} \le L\varepsilon|\rho|^{e_o} \qquad \|\varphi_{i,e+1} - \varphi_{ie}\|_{G\cap U_i,\rho} \le L_o\varepsilon|\rho|^{e_o}$$

$$\|\varphi_e - \varphi_{e_o}\|_{G,\rho} \le L\varepsilon|\rho|^{e_o},$$

then there exists a homomorphism φ_{e+1} over $G\times S_{e+1}$ extending φ_e and satisfying the estimate

$$\|\varphi_{e+1} - \varphi_e\|_{G,\rho} \le L_o'\varepsilon|\rho|^{e_o},$$

such that the following diagram is commutative

$$\begin{array}{ccccccc} \mathcal{O}^q_{e+1} & \xrightarrow{\ \sigma_{1,e+1}\ } & \mathcal{O}^p_{e+1} & \longrightarrow & \mathcal{F}_{e+1} & \longrightarrow & 0 \\ & & \downarrow{\scriptstyle \varphi_{e+1}} & & \| & & \\ \mathcal{O}^q_{e+1} & \xrightarrow{\ \sigma_{2,e+1}\ } & \mathcal{O}^p_{e+1} & \longrightarrow & \mathcal{F}_{e+1} & \longrightarrow & 0 . \end{array}$$

Remarks: 1) The domain G needs no shrinking.

2) The estimates are independent of $e\ge e_o$.

Proof: a) Restricting the homomorphisms of the e_o-situation to X we obtain the diagram

$$\begin{array}{ccccccc} \mathcal{O}^q_X & \xrightarrow{\ \sigma_{1o}\ } & \mathcal{O}^p_X & \longrightarrow & \mathcal{F}_o & \longrightarrow & 0 \\ & & {\scriptstyle \varphi}\downarrow\downarrow{\scriptstyle \varphi_{io}} & & \| & & \\ \mathcal{O}^q_X & \xrightarrow{\ \sigma_{2o}\ } & \mathcal{O}^p_X & \longrightarrow & \mathcal{F}_o & \longrightarrow & 0 . \end{array}$$

Let $U_i' \subset\subset V_i \subset\subset U_i$ define an intermediate covering. For any $\xi \in Z^1(G\cap\mathfrak{V}, \operatorname{Im}\sigma_{2o})$ we find a cochain $(g_i) \in C^o(G\cap\mathcal{U}', \mathcal{O}_X^q)$ such that

$$\xi_{ij} = \sigma_{2o}(g_j - g_i)$$

and

$$\|g_i\|_{G\cap U_i'} \leq K_o\|\xi\|_{G\cap\mathfrak{V}} ,$$

where K_o is the pullback constant of (6.7).

b) Let M_o denote the maximum of the norms of all homomorphisms of the e_o- and O-situation with respect to G or $G\cap U_i$, let $L_o>0$ be arbitrary and choose L_o' such that

$$K_o(3 + L_o + L_oM_o + 2K_oM_o(3 + L_o + L_oM_o)) \leq L_o' .$$

By (7.5), (I) there exists a triangular condition T such that for $(L,L^*,\varepsilon,\theta) \in T'$ we find a homomorphism g_{ie} on $G\cap V_i\times S_e$ extending g_{ie_o} such that $\varphi_e - \varphi_{ie} = \sigma_{2e}g_{ie}$ and

$$\|g_{ie} - g_{ie_o}\|_{G\cap V_i,\rho} \leq L^*\varepsilon|\rho|^{e_o} .$$

We can choose L^* minimal after we fixed L, such that L^* only depends on L.

c) Let $\varepsilon>0$ in addition satisfy

$$2\varepsilon(L + L^* + LL^*) \leq 1$$

and choose further $\theta_\varepsilon > 0$ so small such that for $0 < \theta \leq \theta_\varepsilon$ we have

$\|\sigma_{2e_o} - \sigma_{2o}\|_{G,\rho} \leq \varepsilon$

$\|g_{ie_o} - g_{io}\|_{G\cap U_i,\rho} \leq \varepsilon$

$\|red_e \sigma_{2e_o} g_{ie_o} - red_{e_o} \sigma_{2e_o} g_{ie_o}\|_{G\cap U_i,\sigma} \leq \varepsilon |\rho|^{e_o}$,

cf. (5.1) and (5.7.2), the last estimate being independent $e \geq e_o$. All the above conditions yield a triangular condition E for ε, θ. Let now $(\varepsilon, \theta) \in E$.

d) By (5.8) we obtain the estimate

$$\| red^{e+1} \sigma_{2e} g_{ie}\|_{G\cap V_i,\rho} \leq 3\varepsilon |\rho|^{e_o}.$$

If we define the coefficients $a_{i\nu}$ by

$\Sigma\, a_{i\nu} t^{\nu} = red^{e+1}(\varphi_e - \varphi_{ie} - \sigma_{2e} g_{ie})$,

where summation is over all $\nu \in \Delta_{e+1}$, we have

$$\rho^{\nu} \|a_{i\nu}\|_{G\cap V_i} \leq \gamma_{\nu}\, 3\varepsilon |\rho|^{e_o}$$

for any $\nu \in \Delta_{e+1}$.

e) Let now $\sigma_{k,e+1} = \sigma_{ke} + \Sigma\, \sigma_{k\nu} t^{\nu}$ and $\varphi_{i,e+1} = \varphi_{ie} + \Sigma\, \varphi_{i\nu} t^{\nu}$ be given as in the Lemma and $\rho^{\nu} \|\sigma_{k\nu}\|_{G,\rho}, \rho^{\nu} \|\varphi_{i\nu}\|_{G\cap U_i} \leq \gamma_{\nu} L_o \varepsilon |\rho|^{e_o}$. We have to construct coefficients φ_{ν} defined over G such that mod $\tilde{\mathfrak{J}}_{e+1}$

$$\varphi_e + \Sigma \varphi_{\nu} t^{\nu} - \varphi_{ie} - \Sigma \varphi_{i\nu} t^{\nu} \equiv (\sigma_{2e} + \Sigma \sigma_{2\nu} t^{\nu})(g_{ie} + \Sigma g_{i\nu} t^{\nu})$$

where we also need auxiliary coefficients $g_{i\nu}$. This equation is again equivalent to

$$a_{i\nu} + \varphi_{\nu} - \varphi_{i\nu} = \sigma_{2o} g_{i\nu} + \sigma_{2\nu} g_{io}$$

for $\nu \in \Delta_{e+1}$. Let us define

$$\xi_{ij\nu} = (a_{j\nu} - a_{i\nu}) - (\varphi_{j\nu} - \varphi_{i\nu}) - \sigma_{2\nu}(g_{jo} - g_{io}) \ .$$

Since we can find algebraic solutions over $G \cap U_i$ which can be developed into reduced series over $G' \cap U_i'$ for any $G' \subset\subset G$, $U_i' \subset\subset V_i$, we conclude as in (8.1), f) that $\xi_\nu = (\xi_{ij\nu}) \in Z^1(G \cap \mathcal{U}, \operatorname{Im} \sigma_{2o})$. Moreover we have

$$\rho^\nu \| \xi_\nu \|_{G \cap \mathcal{U}} \leq 2(3 + L_o + L_o M_o) \gamma_\nu \ \varepsilon |\rho|^{e_o} \ .$$

f) By a) we can find $g_{i\nu}$ over $G \cap U_i'$ with $\xi_{ij\nu} = \sigma_{2o}(g_{j\nu} - g_{i\nu})$ and

$$\rho^\nu \| g_{i\nu} \|_{G \cap U_i'} \leq 2K_o(3 + L_o + L_o M_o) \gamma_\nu \ \varepsilon |\rho|^{e_o} \ .$$

By $\varphi_\nu = \varphi_{i\nu} - a_{i\nu} + \sigma_{2\nu} g_{io} + \sigma_{2o} g_{i\nu}$ we now obtain a global section in $\Gamma(G, \mathcal{H}om(\mathcal{O}_X^q, \mathcal{O}_X^p))$ with the estimate

$$\begin{aligned} \rho^\nu \| \varphi_\nu \|_G &\leq K_o \rho^\nu \max_i \| \varphi_{i\nu} - a_{i\nu} + \sigma_{2\nu} g_{io} + \sigma_{2o} g_{i\nu} \|_{G \cap U_i'} \\ &\leq \gamma_\nu K_o (L_o + 3 + L_o M_o + 2K_o M_o (3 + L_o + L_o M_o)) \ \varepsilon |\rho|^{e_o} \\ &\leq \gamma_\nu L_o' \varepsilon |\rho|^{e_o} \ . \end{aligned}$$

If we define $\varphi_{e+1} = \varphi_e + \Sigma \varphi_\nu t^\nu$ we obtain by our construction

$$\varphi_{e+1} - \varphi_{i,e+1} = \sigma_{2,e+1} g_{i,e+1}$$

$$\| \varphi_{e+1} - \varphi_e \|_{G,\rho} \leq L_o' \varepsilon |\rho|^{e_o} \ .$$

This completes the proof of (8.3) .

§ 9 Elimination of obstructions.

Besides the auxiliary Lemmas in § 7, § 8 concerning amalgamation and extensions of homomorphisms with bounds we also need bounds for the process of elimination of the obstructions for extending deformations from S_e to S_{e+1}, cf. (3.4.2), (3.4.3). To do this we have to consider norms for matrices with respect to several charts.

(9.1) Let $\mathbb{U}_i = (U_i, Y_i, j_i, \Omega_i, G_i)$ be a family of special charts on X, where $G_i \subset \Omega_i$ is Stein and $U_i = j_i^{-1}G_i$. By a Lemma of Douady, cf. [6], we can find Stein domains $G_i^j \subset G_i$ such that $U_{ij} = j_i^{-1}G_i^j$. Whenever we deal with such a family of special charts $\mathbb{U}_i$ we will fix the system of the G_i^j. Thus as a special cover we define a pair $\mathfrak{U} = ((\mathbb{U}_i), (G_i^j))$ consisting of a system of charts U_i and a system of sets G_i^j as above. If $\mathfrak{U}'$ and $\mathfrak{U}$ are two special covers, we write $\mathfrak{U}' \subset\subset \mathfrak{U}$, if they have the same index set and if $G_i' \subset\subset G_i$ and $G_i'^j \subset\subset G_i^j$.

(9.2) Let now $\mathfrak{U}$ be a special cover and let $f \in \Gamma(U_\alpha, \mathcal{O}_X)$ where $\alpha = (i_o,\ldots,i_p)$. By abuse of notation we define $\|f\|_{U_\alpha}^{i_\nu}$ to be the infimum of

$$\{\|\tilde{f}\|_{G_{i_\nu}^{i_o}\cap\ldots\cap G_{i_\nu}^{i_p}},\ \tilde{f} \in \Gamma(G_{i_\nu}^{i_o}\cap\ldots\cap G_{i_\nu}^{i_p}, \mathcal{O}_{\Omega_{i_\nu}}),\ j_{i_\nu}^*\tilde{f} = f\}$$

and we put

$$\|f\|_{U_\alpha}^* = \max_\nu \|f\|_{U_\alpha}^{i_\nu} .$$

Finally by $\mathcal{B}^*(U_\alpha)$ we denote the Banach space of all $f \in \Gamma(U_\alpha, \mathcal{O}_X)$ with $\|f\|^*_{U_\alpha} < \infty$. If $A \in \mathrm{Hom}(U_\alpha, \mathcal{O}^q_X, \mathcal{O}^r_X)$ we similarly to (6.1), (6.2) define $\|A\|^{i_\nu}_{U_\alpha}$ as the operator norm and then put

$$\|A\|^*_{U_\alpha} = \max_\nu \|A\|^{i_\nu}_{U_\alpha} .$$

By this definition we again obtain the law

$$\|AB\|^*_{U_\alpha} \le \|A\|^*_{U_\alpha} \|B\|^*_{U_\alpha}$$

for the product of two matrices on U_α. As a consequence of (6.3) we obtain:

<u>(9.3) Lemma:</u> Let $\mathcal{U}' \subset\subset \mathcal{U}$ be finite special covers as above and q, r be given. Then there is a constant K>0 such that for any $\alpha = (i_o \ldots i_p)$ and $A \in \mathrm{Hom}(U_\alpha, \mathcal{O}^q_X, \mathcal{O}^r_X)$

$$\|A|U'_\alpha\|^*_{U'_\alpha} \le K\|A\|^{i_\nu}_{U_\alpha}$$

for any $1 \le \nu \le p$.

(9.4) As in the previous paragraphes let a sequence $\mathfrak{J}_d \subset \mathbb{C}_d\langle t\rangle$, $t = (t_1, \ldots, t_m)$, $d \le e$, be given such that $\mathfrak{J}_{d+1}|\mathbb{C}_d\langle t\rangle = \mathfrak{J}_d$. We assume that $\Lambda(\mathfrak{J}_d) = \Lambda(\mathfrak{J}_{e_o}) = \Lambda$ for $e_o \le d \le e$, where $e_o < e$, and that there exists an ideal $\mathfrak{J}_* \subset \mathbb{C}_{e+1}\langle t\rangle$ with $\mathfrak{J}_*|\mathbb{C}_e\langle t\rangle = \mathfrak{J}_e$ and $\Lambda(\mathfrak{J}_*) = \Lambda$, cf. (5.6.1). We then use the abbreviation red^b_{e+1} for $\mathrm{red}_{\mathfrak{J}_{e+1}}(b)$.

(9.4.1) By S_d we denote the germ of $\mathfrak{J}_d$ and we assume that a coherent S_e-flat sheaf $\mathcal{F}_e$ on $X \times S_e$ is given with Supp $\mathcal{F}_e$

compact. Let $\mathcal{F}_o = \mathcal{F}_e|X$ and let moreover $\mathfrak{U}$ be a finite special cover as above, covering Supp $\mathcal{F}_o$.

(9.4.2) In addition to the above assumptions we suppose that $\mathcal{F}_e$ has a resolution by sheaf systems with respect to $\mathfrak{U} \times S_e$ as described in § 2. We write $\alpha = (i)$, $\beta = (ij)$, $\gamma = (ijk)$ and denote as in §§2,3 by A^e_α, A^e_β, A^e_γ, B^e_α, B^e_β, B^e_γ, $\Phi^e_{\alpha\beta}$, $\Phi^e_{\beta\gamma}$, $\Phi^e_{\alpha\gamma}$, $\Psi^e_{\alpha\beta}$, $\Psi^e_{\beta\gamma}$, $\Psi^e_{\alpha\gamma}$ the first matrices of the system, where we write A^e_α instead of $A_{\alpha S_e}$ etc.

(9.4.3) We assume moreover that all the above matrices have finite norms $\| \quad \|^*_{U_*,\rho}$ and that they are Λ-e-reduced. With respect to the Banach spaces $\mathfrak{B}^*(U_*)$ we can apply the operator red^b_{e+1} and therefore define the matrices $R^{b\nu}_*$ by the following equations, cf. (3.3):

$$\mathrm{red}^b_{e+1} A^e_\alpha B^e_\alpha = \sum_\nu R^{b\nu}_\alpha t^\nu$$

$$\mathrm{red}^b_{e+1} (\Phi^e_{\alpha\beta} A^e_\beta - A^e_\alpha \Psi^e_{\alpha\beta}) = \sum_\nu R^{b\nu}_{\alpha\beta} t^\nu$$

$$\mathrm{red}^b_{e+1} (\Phi^e_{\alpha\beta} \Phi^e_{\beta\gamma} - \Phi^e_{\alpha\gamma}) = \sum_\nu R^{b\nu}_{\alpha\beta\gamma} t^\nu,$$

where summation is only over $\nu \in \Delta_{e+1}$, since there are no terms of order $\leqq e$.

(9.5) To use (5.6.3) we write by using Grauerts division theorem (5.5)

$$A^{e_o}_\alpha B^{e_o}_\alpha = \sum_{\mu \in \Lambda} Y_{\alpha\mu} \omega^{e_o}_\mu + \overline{Y}_\alpha$$

$$\Phi_{\alpha\beta}^{e_o} A_\beta^{e_o} - A_\alpha^{e_o} \Psi_{\alpha\beta}^{e_o} = \sum_{\mu\in\Lambda} Y_{\alpha\beta,\mu} \omega_\mu^{e_o} + \overline{Y}_{\alpha\beta}$$

$$\Phi_{\alpha\beta}^{e_o} \Phi_{\beta\gamma}^{e_o} - \Phi_{\alpha\gamma}^{e_o} = \sum_{\mu\in\Lambda} Y_{\alpha\beta\gamma,\mu} \omega_\mu^{e_o} + \overline{Y}_{\alpha\beta\gamma}$$

where the Y are bounded with respect to the norms $\| \; \|^*_{U_*\,\rho}$ and the $\overline{Y}$ are bounded and reduced. Let $c_{\mu\lambda}$ be defined as in (5.6) for $\mu \in \Lambda$ and $\lambda \in \Lambda_o$. If we define for $\lambda \in \Lambda_o$

$$Z_{*\lambda} = \sum_{\mu\in\Lambda} Y_{*\mu}(0) c_{\mu\lambda}(0)$$

where $*$ substitutes α , $\alpha\beta$ or $\alpha\beta\gamma$, we obtain by (5.6.3) that

$$R_*^{o\nu} - R_*^{b\nu} = \sum_{\lambda\in\Lambda_o} Z_{*\lambda} b_{\lambda\nu} \; . \tag{1}$$

(9.6) Now as in (3.3) we define $\mathbb{a}^{b\nu}$, $\mathbb{b}^{b\nu}$, $\mathbb{c}^{b\nu}$ by the formulae, $\nu \in \Delta_{e+1}$:

$$\mathbb{a}_i^{b\nu} = - P_i^o R_i^{b\nu}$$

$$\mathbb{b}_{ij}^{b\nu} = P_j^o R_{j,ij}^{b\nu} - P_i^o R_{i,ij}^{b\nu}$$

$$\mathbb{c}_{ijk}^{b} = P_i^o(R_{i,ij,ijk}^{b\nu} - R_{i,ik,ijk}^{b\nu}) + P_j^o(R_{j,jk,ijk}^{b} - R_{j,ij,ijk}^{b\nu}) + P_k^o(R_{k,ik,ijk}^{b\nu} - R_{k,jk,ijk}^{b\nu})$$

where P_α^o are the augmenting epimorphisms of the sheaf system, and thus obtain the elements

$$\mathbb{c}^{b\nu} \oplus \mathbb{b}^{b\nu} \oplus \mathbb{a}^{b\nu} \in \operatorname{Ker} d^2 \subset C^2(\mathcal{U}, \mathcal{F}_o, \mathcal{F}_o)$$

with finite norms. Similarly we define for $\lambda \in \Lambda_o$ the elements

$$\zeta_\lambda = \tilde{\mathbb{c}}_\lambda \oplus \tilde{\mathbb{b}}_\lambda \oplus \tilde{\mathbb{a}}_\lambda \in C^2(\mathcal{U}, \mathcal{F}_o, \mathcal{F}_o)$$

by using $Z_{*\lambda}$ instead of $R_*^{b\nu}$. Then by (1) we obtain the equations

$$\mathbb{c}^{o\nu} \oplus \mathbb{b}^{o\nu} \oplus \mathbb{a}^{o\nu} - \mathbb{c}^{b\nu} \oplus \mathbb{b}^{b\nu} \oplus \mathbb{a}^{b\nu} = \sum_{\lambda \in \Lambda_o} \zeta_\lambda b_{\lambda\nu} \tag{2}$$

for $\nu \in \Delta_{e+1}$. Since this equation holds for every b we conclude that $\zeta_\lambda \in Z^2(\mathcal{U}, \mathcal{F}_o, \mathcal{F}_o) = \operatorname{Ker} d^2$.

(9.7) We define the linear operator

$$H : C^1(\mathcal{U}, \mathcal{F}_o, \mathcal{F}_o) \times \mathbb{C}^{\Lambda_o} \longrightarrow Z^2(\mathcal{U}, \mathcal{F}_o, \mathcal{F}_o)$$

by

$$\xi \oplus \eta \oplus (x_\lambda) \longmapsto d^1(\xi \oplus \eta) + \sum_{\lambda \in \Lambda_o} \zeta_\lambda x_\lambda ,$$

where we use the notations of (3.2). H obviously is continuous and Im H has finite codimension because $\dim \operatorname{Ext}^2(X, \mathcal{F}_o, \mathcal{F}_o) < \infty$. Therefore H has closed range. If $\mathfrak{W} \subset\subset \mathcal{U}$ is a second special cover of $\operatorname{Supp} \mathcal{F}_o$ as in (9.1), by the open mapping theorem there exists a constant K_o with the following property.

If $\mathbb{c} \oplus \mathbb{b} \oplus \mathbb{a}$ is defined by a system R_* as in (9.6) and if $\mathbb{c} \oplus \mathbb{b} \oplus \mathbb{a} \in \operatorname{Im} H$, we can find, using the notations of (3.4.3), elements $\xi = (\xi_{ij})$, $\eta = (\eta_i)$ and (b_λ) with $\xi_{ij} = P_j^o \phi_{j,ij}$, $\eta_i = P_i^o A_i$, where $\phi_{j,ij} \in \operatorname{Hom}(V_{ij}, \mathcal{R}_{ij}^o, \mathcal{R}_j^o)$ and $A_i \in \operatorname{Hom}(V_i, \mathcal{R}_i^1, \mathcal{R}_i^o)$ such that

$$\mathbb{c} \oplus \mathbb{b} \oplus \mathbb{a} = d^1(\xi \oplus \eta) + \sum_{\lambda \in \Lambda_o} \zeta_\lambda b_\lambda$$

and $\|A_i\|_{V_i}$, $\|\phi_{j,ij}\|_{V_{ij}}^*$, $|b_\lambda| \leq K_o M$

if $\|R_\alpha\|_{U_\alpha}^*$, $\|R_{\alpha\beta}\|_{U_\beta}^*$, $\|R_{\alpha\beta\gamma} - R_{\alpha\beta'\gamma}\|_{U_\gamma}^* \leq M$

for all possible indices. Here we have included the intermediate steps of going from the R_* to sections in $\mathcal{F}_o$ and going back to holomorphic matrices via the epimorphisms P_i^o .

(9.8) Assume now that $\mathfrak{J}_e$ has an extension $\mathfrak{J} \subset \mathbb{C}_{e+1}\langle t\rangle$ with germ S and $\Lambda(\mathfrak{J}) = \Lambda$, and assume that $\mathcal{F}_e$ can be extended to $X\times S$ as a deformation. By (5.6.1) there is a system $b' = (b'_{\lambda\nu})$ such that $\mathfrak{J} = \mathfrak{J}_{e+1}(b')$. By (3.3.3) we get the elements

$$\mathbb{c}^{b'\nu} \oplus \mathbb{b}^{b'\nu} \oplus \mathbb{a}^{b'\nu} \in \operatorname{Im} d^1 .$$

By (2) of (9.6) we conclude that then

$$\mathbb{c}^{o\nu} \oplus \mathbb{b}^{o\nu} \oplus \mathbb{a}^{o\nu} \in \operatorname{Im} H,$$

and hence we can find A_i^ν, $\phi_{j,ij}^\nu$ and $b_{\lambda\nu}$ as above such that

$$\mathbb{c}^{o\nu} + \mathbb{b}^{o\nu} + \mathbb{a}^{o\nu} = d^1(\xi^\nu \oplus \eta^\nu) + \sum_{\lambda\in\Lambda_o} \zeta_\lambda b_{\lambda\nu}$$

and $\quad \|A_i^\nu\|_{V_i}$, $\|\phi_{j,ij}^\nu\|^*_{V_{ij}}$, $|b_{\lambda\nu}| \leq K_o M$,

if $\quad \|R_\alpha^{o\nu}\|^*_{U_\alpha}$, $\|R_{\alpha\beta}^{o\nu}\|^*_{U_\beta}$, $\|R_{\alpha\beta\gamma}^{o\nu} - R_{\alpha\beta\gamma}^{o\nu}\|^*_{U_\gamma} \leq M$.

Now $b = (b_{\lambda\nu})$ also defines the matrices $R_*^{b\nu}$ as in (9.4.3), and the elements $\mathbb{c}^{b\nu} \oplus \mathbb{b}^{b\nu} \oplus \mathbb{a}^{b\nu}$ as in (9.6). Using the formula (2) above we get

$$\mathbb{c}^{b\nu} \oplus \mathbb{b}^{b\nu} + \mathbb{a}^{b\nu} = d^1(\xi^\nu \oplus \eta^\nu) .$$

Using then (3.4.2), (3.4.3), the matrices A_i^ν, $\phi_{j,ij}^\nu$ constructed here are part of a system

$$\mathfrak{X}^\nu = (A_\alpha^\nu, A_\beta^\nu, A_\gamma^\nu, B_\alpha^\nu, B_\beta^\nu, B_\gamma^\nu, \Phi_{\alpha\beta}^\nu, \dots, \Psi_{\beta\gamma}^\nu)$$

such that with

$$\tilde{A}_i^{e+1} = A_i^e + \sum_\nu A_i^\nu t^\nu$$

$$\tilde{\Phi}_{j,ij}^{e+1} = \phi_{j,ij}^e + \sum_\nu \phi_{j,ij}^\nu t^\nu$$

$$\tilde{\Phi}_{i,ij}^{e+1} = \phi_{i,ij}^e$$

etc.

an extension $\mathcal{F}_{e+1}$ of $\mathcal{F}_e$ is constructed on $X \times S_{e+1}(b)$ as in (3.4.2), where $S_{e+1}(b)$ is the germ of $\mathcal{J}_{e+1}(b)$. By this procedure we have eliminated the obstructions. Here we have used the notation $\tilde{A}_i^{e+1}$ etc. to indicate that we do not have necessarily a sheaf-system resolution, cf. (3.4.2).

(9.9) For later use we derive estimates of $R_*^{o\nu}$ in terms of previous matrices. Let $\mathcal{X}^e$ denote the system of matrices $(A_\alpha^e, A_\beta^e, B_\alpha^e, \Phi_{\alpha\beta}^e, \Phi_{\beta\gamma}^e, \Psi_{\alpha\beta}^e)$ which are given by the sheaf-system resolution of $\mathcal{F}_e$ and by which the matrices $R_*^{o\nu}$ are defined (9.4.3). We write X_*^e for one of the matrices of $\mathcal{X}^e$, where $*$ stands for α, $\alpha\beta$ or $\beta\gamma$.

Let L, $\varepsilon > o$ be any constants satisfying

$$12\varepsilon\, L(2 + L) \leq 1,$$

let ${}^o\omega_\mu^{e+1}$, $\mu \in \Lambda$ be the GW-family of $\mathcal{J}_{e+1}(0)$ and assume that for $\rho = \theta\rho_o$, $0 < \theta \leq 1$, we have for any $\mu \in \Lambda$:

$$\| {}^o\omega_\mu^{e+1} - t^\mu \|_\rho \leq \frac{1}{2\,\mathrm{card}\Lambda}\, \rho^\mu .$$

Let θ be in addition so small that we have the estimates

$$\|X_*^{e_o} - X_*^o\|^*_{U_*,\rho} \leq \varepsilon$$

$$\|\mathrm{red}^o_{e+1} A_\alpha^{e_o} B_\alpha^{e_o}\|_{U_\alpha,\rho} \leq \varepsilon|\rho|^{e_o}$$

$$\|\mathrm{red}^o_{e+1}(\phi_{\alpha\beta}^{e_o} A_\beta^{e_o} - A_\alpha^{e_o}\psi_{\alpha\beta}^{e_o})\|^*_{U_\beta,\rho} \leq \varepsilon|\rho|^{e_o}$$

$$\|\mathrm{red}^o_{e+1}(\phi_{\alpha\beta}^{e_o}\phi_{\beta\gamma}^{e_o} - \phi_{\alpha\gamma}^{e_o})\|^*_{U_\gamma,\rho} \leq \varepsilon|\rho|^{e_o}$$

independent of $e \geq e_o$. This can be achieved by (5.2.1) and (5.7.1).

(9.9.1) Claim: If under the above assumptions

$\|X_*^e - X_*^{e_o}\|^*_{U_*,\rho} \leq L\,\varepsilon|\rho|^{e_o}$, then for any $\nu \in \Delta_{e+1}$

$$\rho^\nu\|R_\alpha^{o\nu}\|_{U_\alpha} \leq \gamma_\nu\,\varepsilon|\rho|^{e_o}$$

$$\rho^\nu\|R_{\alpha\beta}^{o\nu}\|^*_{U_\beta} \leq \gamma_\nu\,\varepsilon|\rho|^{e_o}$$

$$\rho^\nu\|R^o_{\alpha\beta\gamma} - R^o_{\alpha\beta'\gamma}\|^*_{U_\gamma} \leq \gamma_\nu\,\varepsilon|\rho|^{e_o} .$$

This is a direct consequence of (5.8) by which we obtain, using the auxiliary terms

$$\phi^e_{\alpha\beta}A^e_\beta - \phi^{e_o}_{\alpha\beta}A^{e_o}_\beta , \text{ etc.}$$

the estimates:

$$\|\mathrm{red}^o_{e+1} A^e_\alpha B^e_\alpha\|_{U_\alpha,\rho} \leq \varepsilon|\rho|^{e_o}$$

$$\|\mathrm{red}^o_{e+1}(\phi^e_{\alpha\beta}A^e_\beta - A^e_\alpha\psi^e_{\alpha\beta})\|^*_{U_\beta,\rho} \leq \varepsilon|\rho|^{e_o}$$

$$\|\mathrm{red}^o_{e+1}(\phi^e_{\alpha\beta}\phi^e_{\beta\gamma} - \phi^e_{\alpha\beta'}\phi^e_{\beta'\gamma})\|^*_{U_\gamma,\rho} \leq \varepsilon|\rho|^{e_o} .$$

(9.10) Remark: Assume in addition to the assumption in (9.9) that $L \geq C_*$, K_o, card Λ and $8\varepsilon\,K_oL^2 \leq 1$, where K_o is the pullback constant of the operator H in (9.7) and C_* is as in (5.6.2), and assume that

$$\|\omega_\lambda^e - \omega_\lambda^{e_o}\|_\rho \leq \frac{1}{8L^2} |\rho|^{e_o} \quad \text{for } \lambda \in \Lambda_o ,$$

$$\|\sum_{\lambda\in\Lambda_o} c_{\mu\lambda}\omega_\lambda^{e_o} - t^\mu\|_\rho \leq \frac{1}{8\operatorname{card}\Lambda} \rho^\mu \quad \text{for } \mu \in \Lambda .$$

Then by (5.6.2) the estimate $\|{}^o\omega_\mu^{e+1} - t^\mu\|_\rho \leq \frac{1}{2\operatorname{card}\Lambda} \rho^\mu$ for $\mu \in \Lambda$ in (9.9) follows.

Moreover by (9.8) and (9.9.1) we obtain

$$\|A_i^\nu\|_{V_i} , \|\phi_{j,ij}^\nu\|_{V_{ij}}^* , |b_{\lambda\nu}| \leq \frac{\gamma_\nu}{\rho^\nu} K_o \varepsilon|\rho|^{e_o} .$$

Especially from this estimate for $|b_{\lambda\nu}|$ we obtain

$$\|\sum_\nu b_{\lambda\nu}t^\nu\|_\rho \leq L_o \varepsilon|\rho|^{e_o} \leq \frac{1}{8L^2}|\rho|^{e_o} \leq \frac{1}{8C_*\operatorname{card}\Lambda}|\rho|^{e_o} ,$$

such that also $\|\omega_\lambda^{e+1}(b) - \omega_\lambda^{e_o}\|_\rho \leq \frac{1}{8L^2}|\rho|^{e_o}$ for $\lambda \in \Lambda_o$.

Then by (5.6.2) we obtain

$$\|{}^b\omega_\mu^{e+1} - t^\nu\|_\rho \leq \frac{1}{2\operatorname{card}\Lambda}\rho^\mu$$

for any $\mu \in \Lambda$, where $({}^b\omega_\mu^{e+1})$ is the GW-family of $\mathcal{J}_{e+1}(b)$. Note that only for $\lambda \in \Lambda_o$ we have ${}^b\omega_\lambda^{e+1} = \omega_\lambda^{e+1}(b)$. The last estimate is necessary to continue the construction. Moreover if as in (9.9.1) $\|X_*^e - X_*^{e_o}\|_{U_*,\rho}^* \leq L\,\varepsilon|\rho|^{e_o}$, we also have

$$\|\tilde{A}_i^{e+1} - A_i^{e_o}\|_{V_i,\rho} \leq L\,\varepsilon|\rho|^{e_o}$$

$$\|\tilde{\phi}_{j,ij}^{e+1} - \phi_{j,ij}^{e_o}\|_{V_{ij},\rho}^* \leq L\,\varepsilon|\rho|^{e_o}$$

since $L \geq K_o$.

§ 10 Construction of a versal deformation.

Let $\mathcal{F}_o$ be a coherent analytic sheaf with compact support on a complex space X. We are going to construct a deformation $(\mathcal{F},S)$ of $\mathcal{F}_o$ such that $\mathcal{F}_e = \mathcal{F}/\mathfrak{m}_S^{e+1}\mathcal{F}$ is e-versal for every e. To do this we assume that a formal versal deformation is given and then construct $\mathcal{F}_e$ by induction over e. In each step of this induction we derive bound estimates for all entities involved such that we get a convergent system of matrices and functions which define $\mathcal{F}$ and S.

(10.1) Let $(\mathbb{W}_k)_{k\in K}$ be a finite family of special charts $\mathbb{W}_k = (W_k, W_k^*, j_k, G_k^*, G_k)$ where $W_k = j_k^{-1}G_k$ and $G_k \subset\subset G_k^*$ is a privileged polydomain for $\mathcal{O}_X$ and $\mathcal{F}_o$. We assume that the W_k cover $\mathrm{Supp}\,\mathcal{F}_o$. We construct another finite family of Stein open sets $D_{\mu k\ell}$, $(k,\ell) \in K\times K$, $\mu \in I_{k\ell}$, where $I_{k\ell}$ is a finite index set, such that the following is satisfied.

(i) For any $(k,\ell) \in K\times K$ and $\mu \in I_{k\ell}$ there is a privileged polydomain $G_{\mu k\ell} \subset G_k$ with respect to $\mathcal{O}_X$ and $\mathcal{F}_o$ such that $D_{\mu k\ell} = j_k^{-1}G_{\mu k\ell}$.

(ii) $D_{\mu k\ell} \subset\subset W_\ell$ and $D_{\mu k\ell} \subset W_k \cap W_\ell$

(iii) $W_k \cap W_\ell$ is covered by the sets $D_{\mu k\ell}$, $\mu \in I_{k\ell}$, $D_{\mu\ell k}$, $\mu \in I_{\ell k}$, $D_{\mu kh} \cap D_{\nu\ell h}$, $\mu \in I_{kh}$, $\nu \in I_{\ell h}$, $h \in K$.

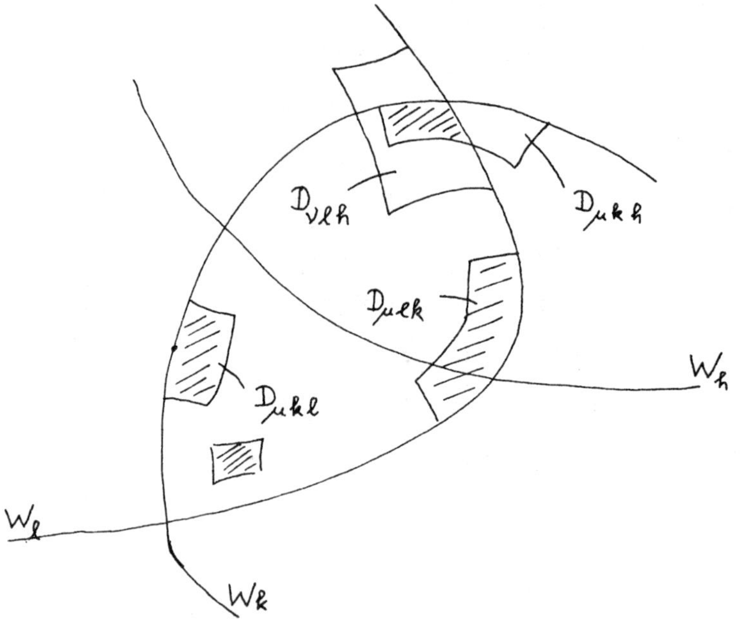

This is done as follows. Let (k,ℓ) be given. If $x \in W_k \cap W_\ell$ let $D_{xk\ell} = j_k^{-1} G_{xk\ell}$ where $G_{xk\ell} \subset\subset G_k$ is any privileged polydomain. If $x \in \partial W_k \cap W_\ell$ let $G_{xk\ell} \subset G_k$ be a privileged polydomain such that $j_k(x)$ is in the relative interior of $\overline{G}_{xk\ell}$ with respect to $\overline{G}_k$, and such that $D_{xk\ell} = j_k^{-1} G_{xk\ell} \subset\subset W_\ell$. By the characterization of privileged polydomains, cf. [19], there are such polydomains.

Denote by $D'_{xk\ell}$ the interior of $\overline{D}_{xk\ell}$ in $\overline{W}_k \cap W_\ell$. Since $\partial W_k \cap \partial W_\ell$ is compact and covered by the sets $D'_{xkh} \cap D'_{x\ell h}$, $h \in K$, there are finitely many (x_ν, h_ν) such that

$$\partial W_k \cap \partial W_\ell \subset \bigcup_\nu D'_{x_\nu k h_\nu} \cap D'_{x_\nu \ell h_\nu} =: D'_{k\ell} . \qquad (*)$$

If $y \in \overline{W}_k \cap \overline{W}_\ell \setminus D'_{k\ell}$ then $y \notin \partial W_k \cap \partial W_\ell$ and there exists some x with $y \in D'_{xk\ell}$ or $y \in D'_{x\ell k}$. Again we can find a finite set of points x_μ such that

$$\overline{W}_k \cap \overline{W}_\ell \setminus D'_{k\ell} \subset\subset \bigcup_\mu D'_{x_\mu k\ell} \cap \bigcup_\nu D'_{x_\nu \ell k} \qquad (**).$$

Now we define $I_{k\ell}$ as the finite index set of all x_λ occuring in (*) or (**) as first index in $D'_{xk\ell}$. Then obviously (i), (ii),

(iii) are satisfied. Note that we have to distinguish between $D_{\mu k \ell}$ and $D_{\mu \ell k}$.

(10.2) Let $(\mathfrak{U}_i)$ be another finite family of special charts $\mathfrak{U}_i = (U_i, U_i^*, j_i, Q_i^*, Q_i)$ where $Q_i \subset\subset Q_i^*$ are open parallel rectangles in some numberspace, Q_i is privileged for $\mathcal{O}_X$ and $\mathcal{F}_o$ and $U_i = j_i^{-1} Q_i$. Let

$$Q_i' \subset\subset \ldots \subset\subset Q_i^{\nu+1} \subset\subset Q_i^{\nu} \subset\subset \ldots \subset\subset Q_i^1 \subset\subset Q_i$$

be a finite sequence of rectangles each of which is again privileged. We assume that the sets $U_i' = j_i^{-1} Q_i'$ cover Supp $\mathcal{F}_o$.

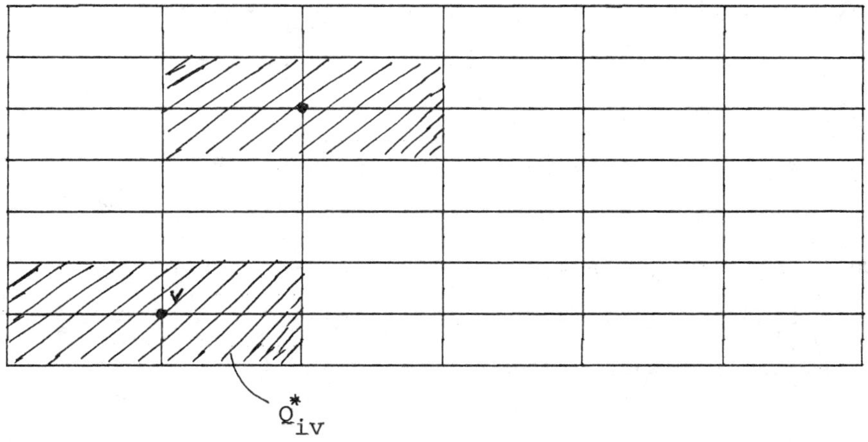

Now we divide Q_i^* into equal sized rectangles and, for any vertex v of this subdivision, $v \notin \partial Q_i^*$, let Q_{iv}^* be as indicated. In the same way we define Q_{iv}' and Q_{iv}^{ν} by the same number of subdivisions. We set $U_{iv}' = j_i^{-1} Q_{iv}', \ldots, U_{iv}^* = j_i^{-1} Q_{iv}^*$. We can assume that the following holds

(i) each W_k is relatively compact in some U_i'

(ii) each U_{iv}^* is relatively compact in some W_k.

We fix a function $\nu \to k$ such that $U^*_{i\nu} \subset\subset W_k$ and then write by abuse of notation $U^*_{(i)k} = U^*_{i\nu}$ if $\nu \to k$.

In order to handle norms with respect to several charts $\mathfrak{U}_i$, cf. (9.1), we can fix Stein domains

$$ {}^{\nu}G^j_i \subset Q^{\nu}_i \quad \text{such that} \quad {}^{\nu+1}G^j_i \subset\subset {}^{\nu}G^j_i $$

and

$$ j_i^{-1}\,{}^{\nu}G^j_i = U^{\nu}_i \cap U^{\nu}_j . $$

Thus we obtain a sequence of special covers of Supp $\mathcal{F}_o$

$$ \mathfrak{U}' = \mathfrak{U}^7 \subset\subset \ldots \subset\subset \mathfrak{U}^1 \subset\subset \mathfrak{U} . $$

(10.3) Because Supp $\mathcal{F}_o$ is compact there exists a formal versal deformation, cf. § 11 or [21], (1.8), which is given by a sequence $(\mathcal{F}'_e, S'_e)$ or e-versal deformations of order e, such that the following holds:

There is an ideal $\mathcal{J}' \subset \mathbb{C}\{t_1,\ldots,t_m\}$ in the formal power series ring such that S'_e is the germ of $\mathcal{J}'_e = \mathcal{J}' + \mathfrak{m}^{e+1}/\mathfrak{m}^{e+1}$ and such that $\mathcal{F}'_{e+1}|X\times S'_e = \mathcal{F}'_e$. Then, cf. [6], there exists $e_o \in \mathbb{N}$ such that $\Lambda(\mathcal{J}'_e) = \Lambda(\mathcal{J}'_{e_o}) = \Lambda$ for $e \geq e_o$. Let Δ be the set of reduced indices with respect to Λ, cf. § 5.

By $\omega_\mu^{e_o}$, $\mu \in \Lambda$, we denote the GW-family of $\mathcal{J}'_{e_o}$ and as in (5.6) we define the polynomials $c_{\mu\lambda}$, the constant C_* and the set $\Lambda_o \subset \Lambda$. Further let ρ_o be such that for $\mu \in \Lambda$

$$ \| \omega_\mu^{e_o} - t^\mu \|_{\rho_o} \leq \frac{1}{2\,\mathrm{card}\Lambda}\, \rho_o^\mu $$

$$ \| \sum_{\lambda\in\Lambda_o} c_{\mu\nu}\omega_\lambda^{e_o} - t^\mu \|_{\rho_o} \leq \frac{1}{8\,\mathrm{card}\Lambda}\, \rho_o^\mu . $$

As before we use the notation $\rho = \Theta\rho_o$ for $0<\Theta\leq 1$.

(10.4) In the following we construct by induction a sequence $(\mathcal{F}_e, S_e)$, $e \geq e_o$, of deformations of $\mathcal{F}_o$ such that each $(\mathcal{F}_e, S_e)$ is e-versal, $(\mathcal{F}_{e_o}, S_{e_o}) = (\mathcal{F}'_{e_o}, S'_{e_o})$ and $\mathcal{F}_e | X \times S_{e-1} \cong \mathcal{F}_{e-1}$ and $\Lambda(S_e) = \Lambda$. Moreover this sequence converges in the following sense.

Denote by ω^e_μ, $\mu \in \Lambda$, the GW-family of the ideal $\mathcal{J}_e$ of S_e and use the same notations as before. Then $\mathcal{F}_e$ is represented by local matrices σ^e_k defined over $W_k \times S_e$ and $\tau^e_{\mu k \ell}$ defined over $D_{\mu k \ell} \times S_e$ as indicated in the diagram

$$\begin{array}{ccccccc}
\mathcal{O}_e^{p''_k} & \xrightarrow[\sigma^e_k]{} & \mathcal{O}_e^{p'_k} & \longrightarrow & \mathcal{F}_e & \longrightarrow & 0 \\
 & & \check{\tau}^e_{\mu k \ell} \Big\downarrow \Big\uparrow \tau^e_{\mu k \ell} & & \Big\| & & \\
\mathcal{O}_e^{p''_\ell} & \xrightarrow[\sigma^e_\ell]{} & \mathcal{O}_e^{p'_\ell} & \longrightarrow & \mathcal{F}_e & \longrightarrow & 0 \quad .
\end{array} \qquad \text{(Dia 1)}$$

These matrices will be bounded and reduced and $\sigma^e_k | S_{e-1} = \sigma^{e-1}_k$, $\tau^e_{\mu k \ell} | S_{e-1} = \tau^{e-1}_{\mu k \ell}$ and we will find constants $0 < L_o < L$ and $\varepsilon, \theta > 0$ such that

$$\| \sigma^e_k - \sigma^{e_o}_k \|_{W_k, \rho} \leq L_o \, \varepsilon |\rho|^{e_o}$$

$$\| \tau^e_{\mu k \ell} - \tau^{e_o}_{\mu k \ell} \|_{D_{\mu k \ell}, \rho} \leq L_o \, \varepsilon |\rho|^{e_o} \qquad \| \omega^e_\lambda - \omega^{e_o}_\lambda \|_\rho \leq \frac{1}{8L^2} |\rho|^{e_o} \quad \text{for } \lambda \in \Lambda_o \, .$$

Here the norm is taken with respect to the privileged chart W_k and $D_{\mu k \ell} \subset W_k$.

(10.4.1) If we have constructed such a sequence of matrices we obtain a deformation $(\mathcal{F}, S)$ as follows. Define

$$\omega_\lambda = \lim_{e\to\infty} \omega_\lambda^e \quad \text{for } \lambda \in \Lambda_o \quad \text{and}$$

$$\sigma_k = \lim_{e\to\infty} \sigma_k^e \quad \text{and } \tau_{\mu k \ell} = \lim_{e\to\infty} \tau_{\mu k \ell}^e .$$

These limits exist by the estimates. The $\omega_\lambda \in \mathbb{C}\langle t\rangle$ define a germ $(S,0)$ in $(\mathbb{C}^m,0)$ and the $\tau_{\mu k\ell}$ induce isomorphisms Coker $\sigma_\ell \longrightarrow$ Coker σ_k which satisfy the collation condition because for any e we obtain such a condition. Also Coker σ_k is S-flat since each Coker σ_k^e is S_e-flat. Hence by collation we obtain a global sheaf $\mathcal{F}$ on $X\times S$ which is a deformation of $\mathcal{F}_o$ and induces $\mathcal{F}_e$ on $X\times S_e$ for any $e \geq e_o$. In § 11 we conclude from this that $(\mathcal{F}, S)$ is versal.

(10.4.2) Induction hypothesis.

Let the assumptions of (10.3) be satisfied and let for fixed $e \geq e_o$ the entities $\mathcal{F}_e$, S_e, $\mathcal{J}_e$, ω_μ^e for $\mu \in \Lambda$, σ_k^e, $\tau_{\mu k\ell}^e$ be given as above such that the above diagrams are commutative with exact rows and such that σ_k^e, $\tau_{\mu k\ell}^e$ are bounded and reduced. Let $0 < L_o < L$ and ε, $\theta > 0$ be constants such that

$$\begin{aligned}
&\| \omega_\mu^e - t^\mu \|_\rho \leq \frac{1}{2\operatorname{card}\Lambda}\, \rho^\mu \quad \text{for } \mu \in \Lambda \\
&\| \omega_\lambda^e - \omega_\lambda^{e_o} \|_\rho \leq \frac{1}{8L^2}\, \rho^\lambda \qquad \text{for } \lambda \in \Lambda_o \\
&\| \sigma_k^e - \sigma_k^{e_o} \|_{W_k,\rho} \leq L_o\, \varepsilon |\rho|^{e_o} \\
&\| \tau_{\mu k\ell}^e - \tau_{\mu k\ell}^{e_o} \|_{D_{\mu k\ell},\rho} \leq L_o\, \varepsilon |\rho|^{e_o}
\end{aligned} \tag{1}$$

Later we will put conditions on L_o, L, ε, θ.

We assume moreover that $\sigma_k^{e_o}$ resp. $\tau_{\mu k\ell}^{e_o}$ are defined over $W_k^* \supset\supset W_k$ resp. $D_{\mu k\ell}^* \supset\supset D_{\mu k\ell}$ and represent the sheaf $\mathcal{F}_{e_o}$ with respect to these bigger coverings in the same way as above.

(10.5) Amalgamation I of the $\tau^{e}_{\mu k\ell}$.

We are going to construct homomorphisms $\Sigma^{e}_{(i)k}$ over $U_{(i)k} \times S_e$ and $T^{e}_{(i)k\ell}$ over $U_{(i)k} \cap U_{(i)\ell} \times S_e$ such that the following diagrams are commutative with exact rows

$$\begin{array}{ccccccc}
\mathcal{O}_e^{p''_k} & \xrightarrow{\Sigma^{e}_{(i)k}} & \mathcal{O}_e^{p'_k} & \longrightarrow & \mathcal{F}_e & \longrightarrow & 0 \\
 & & \uparrow{\scriptstyle T^{e}_{(i)k\ell}} & & \| & & \\
\mathcal{O}_e^{p''_\ell} & \xrightarrow{\Sigma^{e}_{(i)\ell}} & \mathcal{O}_e^{p'_\ell} & \longrightarrow & \mathcal{F}_e & \longrightarrow & 0 .
\end{array} \qquad \text{(Dia 2)}$$

To do this we can assume that $T^{e_o}_{(i)k\ell}$ and $\Sigma^{e_o}_{(i)k}$ are already constructed in the same way as we will do on the bigger sets $U^*_{(i)k\ell}$ and $T^*_{(i)k}$, and such that $T^{e_o}_{(i)k\ell}$ and $\Sigma^{e_o}_{(i)k}$ are bounded and reduced. We define

$$\Sigma^{e}_{(i)k} = \sigma^{e}_{k} | U_{(i)k} \times S_e$$

(10.5.1) There is a triangular condition T_1 such that for $(L_1, \varepsilon, \theta)$ with $(L_o, L_1, \varepsilon, \theta) \in T_1$ and by (1) there are matrices $T^{e}_{(i)k\ell}$ defined over $U_{(i)k\ell} \times S_e$, bounded, reduced, and satisfying (Dia 2) and the estimate

$$\| T^{e}_{(i)k\ell} - T^{e_o}_{(i)k\ell} \|^{i}_{U_{(i)k\ell}, \rho} \leq L_1 \ \varepsilon |\rho|^{e_o} \qquad (2)$$

independent of $e \geqslant e_o$. Here $\| \ \ \|^{i}$ indicates that the norm is taken with respect to the chart $\mathbb{U}_i$. Since $U_{(i)k} \subset\subset W_k$ we can assume by (6.3) that also

$$\| \Sigma^{e}_{(i)k} - \Sigma^{e_o}_{(i)k} \|^{i}_{U_{(i)k}, \rho} \leq L_1 \ \varepsilon |\rho|^{e_o} .$$

Proof: a) Consider first the homomorphism

$$(\sigma_k^e\ ,\ \tau_{\mu k\ell}^e)\ :\ \mathcal{O}_e^{p_k''} \oplus \mathcal{O}_e^{p_\ell'} \longrightarrow \mathcal{O}_e^{p_k'}\ ,$$

which is surjective. Since $D_{\mu k\ell}$ is privileged for $\mathcal{O}_X$ we can find by (7.1) a triangular condition T_o such that for $(L_o, L_o', \varepsilon, \theta) \in T_o$ the following holds. There are homomorphisms $\check{\tau}_{\mu k\ell}^e$ and g^e, cf. (4.1), b), such that

$$\tau_{\mu k\ell}^e\ \check{\tau}_{\mu k\ell}^e + \sigma_k^e g^e = \mathrm{id}\ .$$

and such that

$$\|\check{\tau}_{\mu k\ell}^e - \check{\tau}_{\mu k\ell}^{e_o}\|_{D_{\mu k\ell},\rho} \leq L_o'\ \varepsilon|\rho|^{e_o}\ .$$

We assume here again that $\check{\tau}_{\mu k\ell}^{e_o}$ has been constructed already over $D_{\mu k\ell}^*$, cf. (Dia 1).

b) Now we fix i and (k,ℓ) and let D_α denote one of the sets $D_{\mu k\ell}$, $D_{\mu\ell k}$ or $D_{\mu kh} \cap D_{\nu\ell h}$ such that we obtain a Stein covering (D_α) of $W_{k\ell}$. Let then τ_α^e denote $\tau_{\mu k\ell}^e$, $\check{\tau}_{\mu\ell k}^e$ or $\tau_{\mu kh}^e\ \check{\tau}_{\nu\ell h}^e$ respectively. Let $D_\alpha' \subset\subset D_\alpha$ and (D_α') still cover $U_{(i)k\ell} \subset\subset W_{k\ell}$. By using the method of product estimates as in (5.8) and by using (6.3) for the coefficients of the series we find an estimate

$$\|\tau_\alpha^e - \tau_\alpha^{e_o}\|^i_{D_\alpha' \cap U_i,\rho} \leq L_o''\ \varepsilon|\rho|^{e_o}\ ,$$

where L_o'' depends only on L_o' and the geometric situation.

c) Now we use (7.4), (II) for the covering $D_\alpha' \cap U_i$ of $U_{(i)k\ell}$ and amalgamation I, (4.1), to get a new triangular condition T_o'

such that for $(L_o'', L_1, \varepsilon, \theta) \in T_o'$ and the previous estimates we can find homomorphisms $T^e_{(i)k\ell}$ over $U_{(i)k\ell} \times S_e$ such that (Dia 2) is fulfilled and

$$\| T^e_{(i)k\ell} - T^{e_o}_{(i)k\ell} \|^i_{U_{(i)k\ell}, \rho} \leq L_1 \, \varepsilon |\rho|^{e_o} .$$

Taking the composition of the triangular conditions T_o and T_o' we obtain the desired triangular condition T_1 .

(10.6) <u>Amalgamation of the Σ^e and T^e</u> .

We are going to construct homomorphisms a_i^e over $U_i^1 \times S_e$ and $\Gamma^e_{(i)k}$, $\check{\Gamma}^e_{(i)k}$ over $U^1_{(i)k} \times S_e$ inducing the identity for $\mathcal{F}_e$ such that we have commutative diagrams with exact rows

$$\begin{array}{ccccccc} \mathcal{O}_e^{q_i} & \xrightarrow{a_i^e} & \mathcal{O}_e^{p_i} & \longrightarrow & \mathcal{F}_e & \longrightarrow & 0 \\ & & \check{\Gamma}^e_{(i)k} \downarrow \uparrow \Gamma^e_{(i)k} & & \| & & \\ \mathcal{O}_e^{p_k''} & \xrightarrow{\Sigma^e_{(i)k}} & \mathcal{O}_e^{p_k'} & \longrightarrow & \mathcal{F}_e & \longrightarrow & 0 \end{array} \quad \text{(Dia 3)}$$

As before we assume that the desired homomorphisms of order e_o are already constructed over U_i^* resp. $U^*_{(i)k}$ and are bounded and reduced. The above homomorphisms are constructed with bounds as follows.

(10.6.1) There is a triangular condition T_2 depending only on the e_o-situation such that for any $(L_2, \varepsilon, \theta)$ with $(L_1, L_2, \varepsilon, \theta) \in T_2$ and (2) homomorphisms a_i^e , $\Gamma^e_{(i)k}$, $\check{\Gamma}^e_{(i)k}$ fulfilling (Dia 3) exist such that independent of $e \geq e_o$:

$$\|a_i^e - a_i^{e_o}\|^i_{U^1_i,\rho} \le L_2\, \varepsilon |\rho|^{e_o}$$

$$\|\Gamma^e_{(i)k} - \Gamma^{e_o}_{(i)k}\|^i_{U^1_{(i)},\rho} \le L_2\, \varepsilon |\rho|^{e_o} \tag{3}$$

$$\|\check{\Gamma}^e_{(i)k} - \check{\Gamma}^{e_o}_{(i)k}\|^i_{U^1_{(i)k},\rho} \le L_2\, \varepsilon |\rho|^{e_o} .$$

Proof: a) We use amalgamation II and the main Lemma (7.4). In the process of amalgamation we investigate the following typical situation

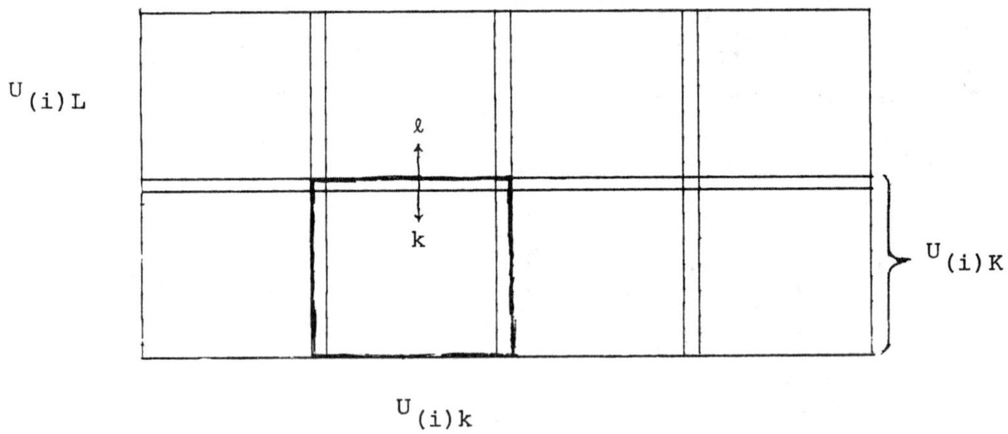

Consider two strips of "rectangles" (intersected with X) $U_{(i)K} = \bigcup U_{(i)k}$ and $U_{(i)L} = \bigcup U_{(i)\ell}$ such that there is a bijection $L \longleftrightarrow K$ such that $U_{(i)\ell}$ and $U_{(i)k}$ are amalgamatable if $\ell \longleftrightarrow k$, and such that $U_{(i)L}$ and $U_{(i)K}$ are amalgamatable. We suppress the index i for this proof now, since it is fixed.

b) First we describe the algebraic process of construction without shrinking.

(1) Assume that a^e_K on U_K, a^e_L on U_L, $\Gamma^e_{K,k}$, $\check{\Gamma}^e_{K,k}$ over U_k, and $\Gamma^e_{L,\ell}$, $\check{\Gamma}^e_{L,\ell}$ over U_ℓ have been constructed with the diagrams

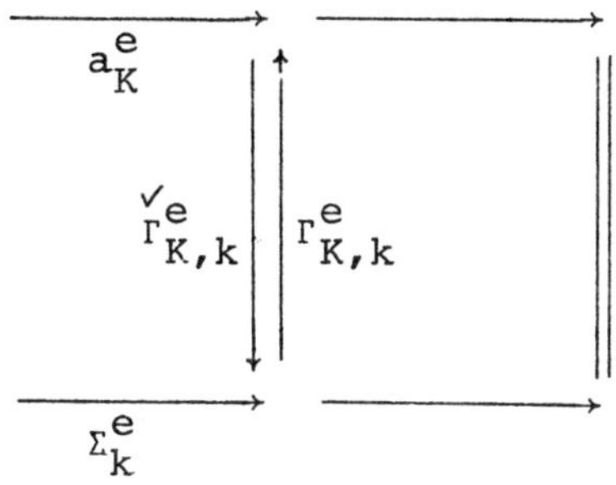

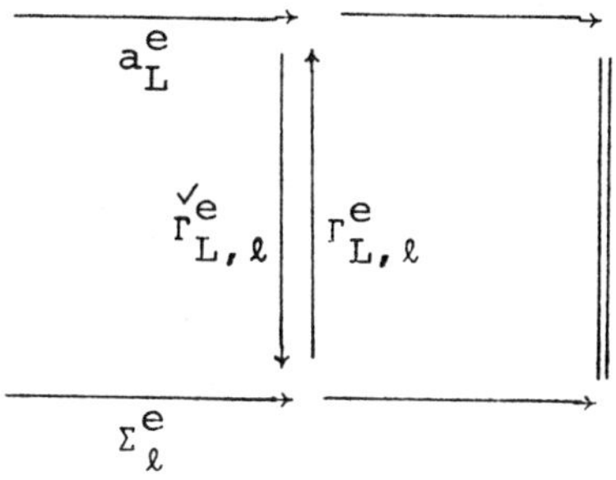

(2) For $k \longleftrightarrow \ell$ we define $\varphi^e_{KL,k\ell} = \Gamma^e_{K,k}\,\check{\Gamma}^e_{L,\ell}$. Then the induced isomorphism is the identity for $\mathcal{F}_e$. By amalgamation I, (4.1) we find a homomorphism φ^e_{KL} on $U_{KL} = U_K \cap U_L$ also inducing the identity for $\mathcal{F}_e$ such that we have a diagram

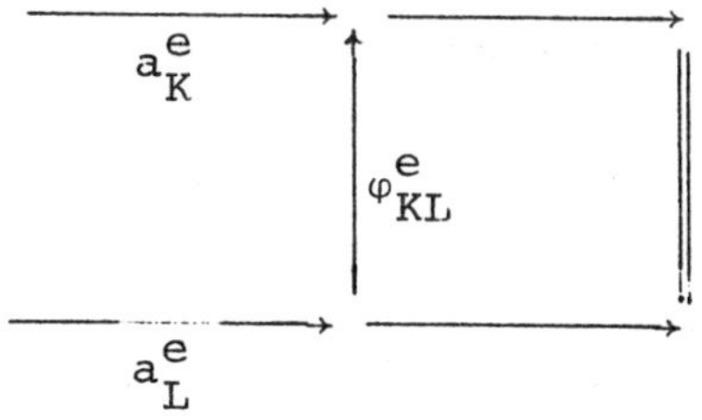

(3) By (7.4), (I) we find homomorphisms ψ^e_{KL} and f^e_{KL} over U_{KL} such that, cf. (4.3),

$$a^e_K\,\psi^e_{KL} = \varphi^e_{KL}\,a^e_L\,,$$

$$a^e_K\,f^e_{KL} = \mathrm{id} - \varphi^e_{KL}\,\varphi^e_{LK}\quad .$$

By the process (4.3) we construct new matrices

$$\hat{\varphi}^e_{KL}\,,\ \hat{\psi}^e_{KL}$$

which are invertible. By Cartan's Lemma, (6.12), we finally obtain on $U_M = U_K \cup U_L$ homomorphisms a^e_M, on U_K resp. U_L homomorphisms Γ^e_{MK}, $\check{\Gamma}^e_{MK}$ resp. Γ^e_{ML}, $\check{\Gamma}^e_{ML}$ with diagrams

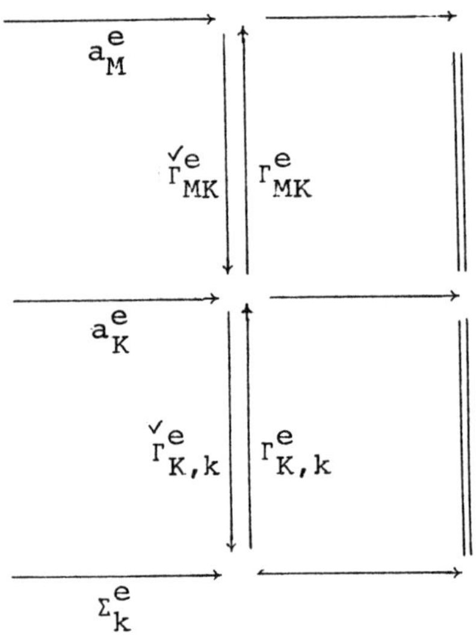

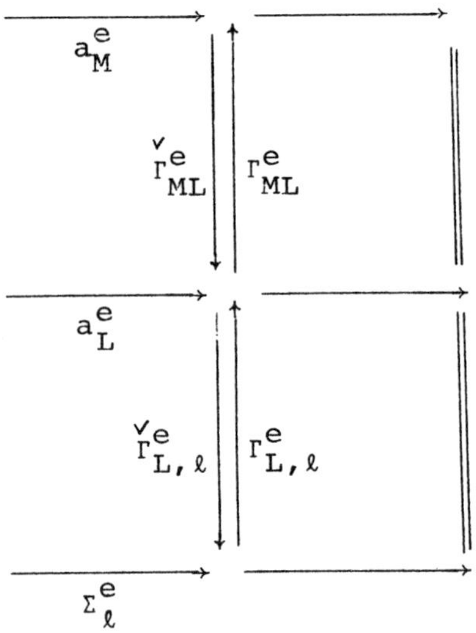

By substituting a global isomorphism for $\mathcal{F}_e$ on U_M we can again assume that $\Gamma^e_{MK}, \ldots$ induce the identity, since by (4.3) $\overline{\Gamma^e_{MK}} = \overline{\Gamma^e_{ML}}$ for the induced isomorphisms.

(4) If we define $\Gamma^e_{M,m}$ by $\Gamma^e_{MK}\Gamma^e_{K,m}$ resp. $\Gamma^e_{ML}\Gamma^e_{L,m}$ for $m \in K$ resp. $m \in L$ we are in the same situation for U_M as for U_K and U_L .

c) To give estimates for steps (1) - (4) we consider a finite number of intermediate "rectangles"

$$U_i^1 \subset\subset \ldots \subset\subset V_i^2 \subset\subset V_i^1 \subset\subset U_i$$

which we subdivide in the same way as U_i, thus obtaining $V^\nu_{(i)k}$ etc. We also assume that all homomorphisms of order e_o in the construction are obtained with respect to U_i^*, $U^*_{(i)k}$ and are bounded and reduced.

d) Now every step of (1) - (4) is covered by a Lemma with bound estimates in the previous sections, and all estimates are of similar type involving triangular conditions. We use (7.4), (I), (II) and Cartan's Lemma (6.12) for the estimates. In every step we obtain a triangular condition accompanied by some shrinking yielding estimates of the type $\|X_*^e - X_*^{e_o}\|_{*\rho}^i \leq L\,\varepsilon|\rho|^{e_o}$. By taking intersections and compositions of the triangular conditions, we obtain a new triangular condition independent of $e \geq e_o$, which yields estimates for the constructed homomorphisms in b). Starting with (10.5) and repeating this process a finite number of times we get the final triangular condition for (10.6.1) with $U_i^1 \subset\subset U_i$.

(10.7) We now construct relating homomorphisms φ_{ij}^e for a_i^e and a_j^e defined over $U_{ij}^2 \times S_e$ such that we have the commutative diagram

$$\begin{array}{ccccccc}
\mathcal{O}_e^{q_j} & \xrightarrow[a_j^e]{} & \mathcal{O}_e^{p_j} & \longrightarrow & \mathcal{F}_e & \longrightarrow & 0 \\
 & & \Big\downarrow \varphi_{ij}^e & & \Big\| & & \\
\mathcal{O}_e^{q_i} & \xrightarrow[a_i^e]{} & \mathcal{O}_e^{p_i} & \longrightarrow & \mathcal{F}_e & \longrightarrow & 0
\end{array} \qquad \text{(Dia 4)}$$

We assume that $\varphi_{ij}^{e_o}$ are constructed in the same way over $U_{ij} \times S_{e_o}$ and are bounded and reduced.

(10.7.1) There is a triangular condition T_3 depending only on the e_o-situation such that for $(L_3, \varepsilon, \theta)$ with $(L_2, L_3, \varepsilon, \theta) \in T_3$

and by the estimates (3) homomorphisms φ^e_{ij} fulfilling (Dia 4) exist such that independent of $e \geq e_o$:

$$\| \varphi^e_{ij} - \varphi^{e_o}_{ij} \|^*_{U^2_{ij}, \rho} \leq L_3 \ \varepsilon |\rho|^{e_o} \qquad (4)$$

Proof: a) We fix i and j and let $U^2_i \subset\subset V'_i \subset\subset V_i \subset\subset U^1_i$ be intermediate rectangles with associated $V'_{(i)k} \subset\subset V_{(i)k}$ as before. Let $U^1_{(ij)k\ell} = U^1_{(i)k} \cap U^1_{(j)\ell}$ etc. and define

$$\varphi^e_{(ij)k\ell} = \Gamma^e_{(i)k} \check{\Gamma}^e_{(j)\ell} \ .$$

These homomorphisms relate a^e_i and a^e_j over $U^1_{(ij)k\ell} \times S_e$. By (6.3) there is a geometrical constant $C>0$ such that

$$\| \check{\Gamma}^e_{(j)\ell} - \check{\Gamma}^{e_o}_{(j)\ell} \|^i_{V_{(ij)k\ell}, \rho} \leq C \| \check{\Gamma}^e_{(j)\ell} - \check{\Gamma}^{e_o}_{(j)\ell} \|^j_{U^1_{(ij)k\ell}, \rho} \ .$$

By (3) we get a new constant L'_2 depending only on L_2 and C such that

$$\| \varphi^e_{(ij)k\ell} - \varphi^{e_o}_{(ij)k\ell} \|^i_{V_{(ij)k\ell}, \rho} \leq L'_2 \ \varepsilon |\rho|^{e_o} \ .$$

b) Now we are in the situation of amalgamation I, (4.1). To obtain bounds we use the covering $\mathfrak{V}_{ij} = (V_{(ij)k\ell})$ of V_{ij} and (7.4), (II) with respect to $Z^1(\mathfrak{V}_{ij} \cap V_{ij} \times S_e, \operatorname{Im} a^e_i)$. Thus we obtain a triangular condition T' such that for $(L'_2, L'_3, \varepsilon, \theta) \in T'$ we can find the desired homomorphisms φ^e_{ij} satisfying (Dia 4) and

$$\| \varphi^e_{ij} - \varphi^{e_o}_{ij} \|^i_{V'_{ij}, \rho} \leq L'_3 \ \varepsilon |\rho|^{e_o} \ .$$

If we use (6.3) again we find a new constant L_3 depending only on L'_3 and the geometrical situation such that

$$\| \varphi^e_{ij} - \varphi^{e_o}_{ij} \|^*_{U^2_{ij}, \rho} \leq L_3 \ \varepsilon |\rho|^{e_o} \ .$$

Modifying the triangular condition with respect to the constants

L_2, L_3 we get the desired result.

(10.8) For later use we construct now relating homomorphisms $T^e_{(k)i}$, $\check{T}^e_{(k)i}$ between σ^e_k and a^e_i on $W_k \cap U^3_i \times S_e$ such that we have the commutative diagram

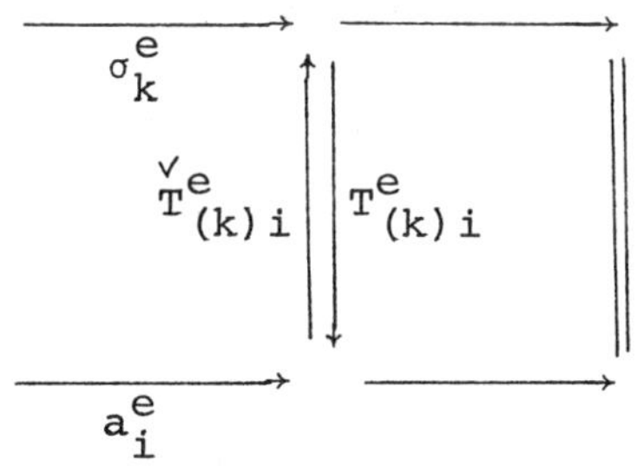

(Dia 5)

We assume that $T^{e_o}_{(k)i}$, $\check{T}^{e_o}_{(k)i}$ are constructed in the same way over $W^*_k \cap U^*_i \times S_{e_o}$ and are bounded and reduced.

(10.8.1) There is a triangular condition T^*_3 depending only on the e_o-situation such that for $(L_2, L^*_3, \varepsilon, \theta) \in T^*_3$ and by the previous estimates homomorphisms $T^e_{(k)i}$, $\check{T}^e_{(k)i}$ exist fulfilling (Dia 5) and satisfying independent of $e \geq e_o$

$$\begin{aligned} \|T^e_{(k)i} - T^{e_o}_{(k)i}\|_{W_k \cap U^3_i, \rho} &\leq L^*_3 \, \varepsilon |\rho|^{e_o} \\ \|\check{T}^e_{(k)i} - \check{T}^{e_o}_{(k)i}\|_{W_k \cap U^3_i, \rho} &\leq L^*_3 \, \varepsilon |\rho|^{e_o} \end{aligned} \qquad (5)$$

where the norm is taken with respect to the privileged chart W_k.

Proof: a) It is easy to see that there is a finite Stein covering $(W_{\kappa\ell})_{\kappa,\ell}$ of $\overline{W}_k \cap \overline{U}^2_i$ such that for every κ, ℓ there is some $\mu \in I_{\kappa\ell}$ with

$$W_k \cap W_{\kappa\ell} \subset D_{\mu\kappa\ell} \cap U^2_{(i)\ell} \quad .$$

On $W_{\kappa\ell}$ we then define

$$T^e_{\kappa\ell} = \operatorname{red}_e \Gamma^e_{(i)\ell} \check{\tau}^e_{\mu\kappa\ell}$$

$$\check{T}^e_{\kappa\ell} = \operatorname{red}_e \tau^e_{\mu\kappa\ell} \check{\Gamma}^e_{(i)\ell} .$$

Then both homomorphisms induce the identity on $\mathcal{F}_e$.
By (10.6.1), (6.3) and by using the product rule decomposition we obtain a constant L_3' depending on L_o, L_2 and the geometrical situation such that

$$\| T^e_{\kappa\ell} - T^{e_o}_{\kappa\ell} \|^k_{W_{\kappa\ell}\cap W_k, \rho} \leq L_3' \ \varepsilon |\rho|^{e_o}$$

$$\| \check{T}^e_{\kappa\ell} - \check{T}^{e_o}_{\kappa\ell} \|^k_{W_{\kappa\ell}\cap W_k, \rho} \leq L_3' \ \varepsilon |\rho|^{e_o}$$

where $\| \ \|^k$ is defined with respect to the privileged chart W_k .
Note that $W_{\kappa\ell}\cap W_k \subset\subset U^1_{(i)\ell}$. Since $W_k \cap U^2_i \subset\subset U^1_i$ we can also assume thet

$$\| a^e_i - a^{e_o}_i \|^k_{W_k \cap U^2_i, \rho} \leq L_3' \ \varepsilon |\rho|^{e_o} .$$

b) Now we use amalgamation I, (4.1) and (7.5), (II) with respect to $\mathfrak{M} \cap W_k \cap U^2_i$ and Im σ^2_k, Im a^e_i without shrinking the privileged set W_k to obtain a triangular condition T_3' such that for $(L_3', L_3^*, \varepsilon, \theta) \in T_3'$ the following holds, cf. proof of (10.7.1):
There are homomorphisms $T^e_{(k)i}$, $\check{T}^e_{(k)i}$ fulfilling (Dia 5) and such that independent of $e \geq e_o$

$$\| T^e_{(k)i} - T^{e_o}_{(k)i} \|^k_{W_k \cap U^3_i, \rho} \leq L_3^* \ \varepsilon |\rho|^{e_o}$$

$$\| \check{T}^e_{(k)i} - \check{T}^{e_o}_{(k)i} \|^k_{W_k \cap U^3_i, \rho} \leq L_3^* \ \varepsilon |\rho|^{e_o} .$$

Finally modifying the triangular condition in the constants L we obtain the desired result.

(10.9) We construct now in addition to a_i^e, φ_{ji}^e homomorphisms b_i^e, c_i^e, ψ_{ji}^e, θ_{ji}^e, f_{kji}^e, g_{kji}^e, h_{kji}^e with respect to $\mathfrak{U}^3 \times S_e$ such that we have the commutative diagrams with exact rows

$$\begin{array}{ccccccccc}
 & \xrightarrow{c_i^e} & & \xrightarrow{b_i^e} & & \xrightarrow{a_i^e} & & \longrightarrow & \\
 & & \Big\downarrow \theta_{ji}^e & & \Big\downarrow \psi_{ji}^e & & \Big\downarrow \varphi_{ji}^e & & \Big\| \\
 & \xrightarrow{c_j^e} & & \xrightarrow{b_j^e} & & \xrightarrow{a_j^e} & & \longrightarrow &
\end{array} \qquad \text{(Dia 6)}$$

and such that we have the relations

$$\left.\begin{array}{l} a_j^e \psi_{ji}^e = \varphi_{ji}^e a_i^e \\ b_j^e \theta_{ji}^e = \psi_{ji}^e b_i^e \end{array}\right\} \quad \text{over } U_{ij}^3 \times S_e$$

$$\left.\begin{array}{l} a_k^e f_{kji}^e = \varphi_{ki}^e - \varphi_{kj}^e \varphi_{ji}^e \\ b_k^e g_{kji}^e = \psi_{ki}^e - \psi_{kj}^e \psi_{ji}^e \\ c_k^e h_{kji}^e = \theta_{ki}^e - \theta_{kj}^e \theta_{ji}^e \end{array}\right\} \quad \text{over } U_{ijk}^3 \times S_e .$$

Again we assume that the corresponding homomorphisms for order e_o have been constructed in the same way with respect to $\mathfrak{U}^*$, and such that they are bounded and reduced.

(10.9.1) There is a triangular condition T_4 such that for $(L_3, L_4, \varepsilon, \theta) \in T_4$ and the estimates (3), (4) we can find

$b_i^e,\ldots,h_{kji}^e$ with respect to $\mathfrak{U}^3 \times S_e$ extending $b_i^{e_o},\ldots,h_{kji}^{e_o}$, satisfying the above relations and independent of $e \geq e_o$ the estimates

$$\left.\begin{array}{l} \|b_i^e - b_i^{e_o}\|_{U_i^1,\rho} \\ \\ \|c_i^e - c_i^{e_o}\|_{U_i^1,\rho} \\ \\ \|\psi_{ji}^e - {}^{e_o}_{ji}\|^*_{U_{ij}^3,\rho} \\ \\ \|\theta_{ji}^e - \theta_{ji}^{e_o}\|^*_{U_{ij}^3,\rho} \\ \\ \|f_{kji}^e - f_{kji}^{e_o}\|^*_{U_{ijk}^3,\rho} \\ \\ \|g_{kji}^e - g_{kji}^{e_o}\|^*_{U_{ijk}^3,\rho} \\ \\ \|h_{kji}^e - h_{kji}^{e_o}\|^*_{U_{ijk}^3,\rho} \end{array}\right\} \leq L_4 \, \varepsilon |\rho|^{e_o} \quad . \tag{6}$$

Proof: a) First since U_i^1 is privileged we can find by (7.3) a triangular condition T_4' such that by (10.6.1), (3) and $(L_2, L_4', \varepsilon, \theta) \in T_4'$ the following holds:

There are homomorphisms b_i^e and c_i^e such that the lines

$$\mathcal{O}_e^{s_i} \xrightarrow[c_i^e]{} \mathcal{O}_e^{r_i} \xrightarrow[b_i^e]{} \mathcal{O}_e^{q_i} \xrightarrow[a_i^e]{} \mathcal{O}_e^{p_i}$$

are exact and such that

$$\|b_i^e - b_i^{e_o}\|_{U_i^1,\rho} \leq L_4' \, \varepsilon |\rho|^{e_o}$$

$$\|c_i^e - c_i^{e_o}\|_{U_i^1,\rho} \leq L_4' \, \varepsilon |\rho|^{e_o}$$

independent of $e \geq e_o$. We can assume that L_4' is minimal for the given L_2 and the condition $L_4' \geq L_3,\ L_2$.

b) Next we use (7.4), (I) and a finite number of intermediate rectangles U_i^3 cc ... cc V_i' cc V_i cc U_i^2 to solve the above equations where b_i^e, c_i^e are known already. Summarizing all steps we get a new triangular condition T_4'' such that for $(L_3, L_4, \varepsilon, \theta) \in T_4''$ we can find $\psi_{ji}^e, \ldots, h_{kji}^e$ extending $\psi_{ji}^{e_o}, \ldots, h_{kji}^{e_o}$ and satisfying the above equations and estimates (6). Finally modifying the conditions on the constants L we get the desired triangular condition.

(10.10) By (10.9) and the construction in § 2 we get now a resolution of $\mathcal{F}_e$ by sheaf systems with respect to the covering $\mathcal{U}^3$. Thus with the notation of § 2 we obtain the matrices A_α^e, B_α^e, $\Phi_{\alpha\beta}^e$, $\Psi_{\alpha\beta}^e, \ldots$. Furthermore $A_\alpha^{e_o}, \ldots, \Psi_{\alpha\beta}^{e_o}$ can be assumed constructed in the same way with respect to $\mathcal{U}^*$ and we can assume that all these matrices for order e_o are bounded and reduced. Since $A_\alpha^e, \ldots, \Psi_{\alpha\beta}^e$ are obtained by simple algebraic operations we can use the product rules, cf. (5.7), (5.8), if necessary, to get a constant L_4^* depending only on L_4 and the e_o-situation, such that by (6) we have

$$\|A_\alpha^e - A_\alpha^{e_o}\|^*_{U^4_\alpha, \rho} \leq L_4^* \,\varepsilon |\rho|^{e_o}$$

$$\vdots \qquad\qquad (6^*)$$

$$\|\Psi_{\alpha\beta} - \Psi_{\alpha\beta}^{e_o}\|^*_{U^4_\beta, \rho} \leq L_4^* \,\varepsilon |\rho|^{e_o} .$$

Here we have to use the next shrinking $\mathcal{U}^4$ cc $\mathcal{U}^3$ since we have to estimate homomorphisms like $f_{kji} a_i$ with respect to several charts and therefore to use (6.3). However by this we only get a multiplication by a geometrical constant.

(10.11) **Let** now $L > L_o, L_1, L_2, L_3, L_3^*, L_4, L_4^*$, card $\Lambda, C_*, K_o, \| X_*^{e_o} \|^*_{U_*^4, \rho}$, where X_*^e is one of the matrices of (10.10) and where K_o is the pullback constant of (9.7). Let (ε, θ) satisfy all the previous triangular conditions, for which we will use the notation $(\varepsilon,\theta) \in E$. Let E be modified such that in addition we have $8\varepsilon K_o L^2 \leq 1$ and for θ, cf. (9.10):

$$\| \omega_\lambda^e - \omega_\lambda^{e_o} \|_\rho \leq \frac{1}{8L^2} |\rho|^{e_o} \quad \text{for } \lambda \in \Lambda_o.$$

Then by (9.10) also

$$\| {}^o\omega_\mu^{e+1} - t^\mu \|_\rho \leq \frac{1}{2\,\text{card}\Lambda} \rho^\mu \quad \text{for } \mu \in \Lambda.$$

If we define $R_\alpha^{o\nu}$, $R_{\alpha\beta}^{o\nu}$, $R_{\alpha\beta\gamma}^{o\nu}$, $\nu \in \Delta_{e+1}$ as in (9.4.3) we obtain by (6^*) and (9.9) the estimates

$$\rho^\nu \| R_\alpha^{o\nu} \|^*_{U_\alpha^4} \leq \gamma_\nu \quad \varepsilon |\rho|^{e_o}$$

$$\rho^\nu \| R_{\alpha\beta}^{o\nu} \|^*_{U_\beta^4} \leq \gamma_\nu \quad \varepsilon |\rho|^{e_o}$$

$$\rho^\nu \| R_{\alpha\beta\gamma}^{o\nu} \|^*_{U_\gamma^4} \leq \gamma_\nu \quad \varepsilon |\rho|^{e_o}.$$

(10.12) As in [6] or [21], (5.9) we conclude by the existence of the formal versal deformation that $(\mathcal{F}_e, S_e)$ is e-versal and that there exists an extension $(\mathcal{F}^*_{e+1}, S^*_{e+1})$ of $(\mathcal{F}_e, S_e)$ with ideal $\mathfrak{J}^*_{e+1} \subset \mathbb{C}_{e+1}\langle t \rangle$

such that $\Lambda(\mathcal{J}^*_{e+1}) = \Lambda$. Now by (9.8), using the same notations, we can find A^ν_i, $\Phi^\nu_{j,ij}$ and $b_{\lambda\nu}$, $\nu \in \Delta_{e+1}$, $\lambda \in \Lambda_o$, defined on U^5_i, U^5_{ij} respectively such that

$$\rho^\nu \|A^\nu_i\|_{U^5_i} \ , \ \rho^\nu \|\Phi^\nu_{j,ij}\|^*_{U^5_{ij}} \ , \ \rho^\nu |b_{\lambda\nu}| \leq \gamma_\nu K_o \ \varepsilon |\rho|^{e_o} \ .$$

Let as in (9.8) for $\lambda \in \Lambda_o$

$$\omega^{e+1}_\lambda(b) \ := \omega^e_\lambda + \sum_\nu b_{\lambda\nu} t^\nu$$

and set $\mathcal{J}_{e+1} = \mathcal{J}_{e+1}(b)$ for the ideal generated by $\omega^{e+1}_\lambda(b)$, $\lambda \in \Lambda_o$. By $8\varepsilon K_o L^2 \leq 1$ we obtain the estimate

$$\|\omega^{e+1}_\lambda(b) - \omega^{e_o}_\lambda\|_\rho \leq \frac{1}{8L^2} |\rho|^{e_o} \quad \text{for } \lambda \in \Lambda_o \ .$$

If ω^{e+1}_μ, $\mu \in \Lambda$, denotes the GW-family of $\mathcal{J}_{e+1}$ we also obtain the estimate, cf. (9.10),

$$\|\omega^{e+1}_\mu - t^\mu\|_\rho \leq \frac{1}{2\operatorname{card}\Lambda} \rho^\mu \quad \text{for } \mu \in \Lambda \ .$$

Let S_{e+1} denote the germ of $\mathcal{J}_{e+1}$.
As in (9.8) we define

$$\tilde{A}^{e+1}_i = A^e_i + \sum_\nu A^\nu_i t^\nu$$

$$\tilde{\Phi}^{e+1}_{j,ij} = \Phi^e_{j,ij} + \sum_\nu \Phi^\nu_{j,ij} t^\nu$$

$$\Phi^{e+1}_{i,ij} = \Phi^e_{i,ij} \ .$$

Using (3.4.2), (3.4.3), cf. (9.8), we also can construct matrices A^ν_β, A^ν_γ, $B^\nu_\alpha, \ldots, \Psi^\nu_{\beta\gamma}$ not necessarily with bounds. Then we also obtain $\tilde{A}^{e+1}_\beta, \ldots, \tilde{\Psi}^{e+1}_{\beta\gamma}$ defined in the same way as $\tilde{A}^{e+1}_i$. By (3.4.2) we obtain by these matrices a sheaf $\mathcal{F}_{e+1}$ on $X \times S_{e+1}$

which extends the deformation $(\mathcal{F}_e, S_e)$. Moreover we can assume that $\tilde{\Phi}^{e+1}_{i,ij}$ and $\tilde{\Phi}^{e+1}_{j,ij}$ induce the identity for $\mathcal{F}_{e+1}$.

(10.13) Setting $\tilde{a}^{e+1}_i = \tilde{A}^{e+1}_i$ and $\tilde{\varphi}^{e+1}_{ji}(x) = \tilde{\Phi}^{e+1}_{j,ij}(x\oplus 0)$ on $U^5_i \times S_{e+1}$ and $U^5_{ij}\times S_{e+1}$ we obtain extensions of a^e_i and φ^e_{ji} and the representation of $\mathcal{F}_{e+1}$

$$\begin{array}{ccccccc}
\mathcal{O}^{q_i}_{e+1} & \xrightarrow[\tilde{a}^{e+1}_i]{} & \mathcal{O}^{p_i}_{e+1} & \longrightarrow & \mathcal{F}_{e+1} & \longrightarrow & 0 \\
 & & \tilde{\Phi}^{e+1}_{ij}\Big\uparrow\Big\downarrow\tilde{\varphi}^{e+1}_{ji} & & \Big\| & & \\
\mathcal{O}^{q_j}_{e+1} & \xrightarrow[\tilde{a}^{e+1}_j]{} & \mathcal{O}^{p_j}_{e+1} & \longrightarrow & \mathcal{F}_{e+1} & \longrightarrow & 0 .
\end{array} \qquad \text{(Dia 7)}$$

Moreover by the construction above we have

$$\begin{aligned}
\rho^{\nu}\|a^{\nu}_i\|_{U^5_i} &\leq \gamma_{\nu} K_o\, \varepsilon|\rho|^{e_o} \\
\rho^{\nu}\|\varphi^{\nu}_{ji}\|^{*}_{U^5_{ij}} &\leq \gamma_{\nu} K_o\, \varepsilon|\rho|^{e_o}
\end{aligned} \qquad (7)$$

if all the above triangular conditions are verified i.e. $(\varepsilon,\theta)\in\mathbf{E}$ Finally note that as in [21], (5.9), the sheaf $\mathcal{F}_{e+1}$ is (e+1)-versal since the reducing system remains the same.

(10.14) To complete the induction process we extend now σ^e_k and $\tau^e_{\mu k \ell}$ without shrinking W_k and $D_{\mu k\ell}$. First we construct σ^{e+1}_k by using Lemma (8.1). Since we have obtained (Dia 5) and (Dia 7) the algebraic assumptions of (8.1) are verified. By (3), (4) and (5) we have the estimates

$$\left.\begin{array}{l}
\|a_i^e - a_i^{e_o}\|^i_{U_i^1,\rho} \leq L_2\ \varepsilon|\rho|^{e_o} \\
\|\varphi_{ij}^e - \varphi_{ij}^{e_o}\|^*_{U^2_{ij},\rho} \leq L_3\ \varepsilon|\rho|^{e_o} \\
\|T^e_{(k)i} - T^{e_o}_{(k)i}\|^k_{W_k\cap U^3_i,\rho} \leq L_3^*\ \varepsilon|\rho|^{e_o} \\
\|\check{T}^e_{(k)i} - \check{T}^{e_o}_{(k)i}\|^k_{W_k\cap U^3_{ij},\rho} \leq L_3^*\ \varepsilon|\rho|^{e_o} \\
\|\sigma_k^e - \sigma_k^{e_o}\|^k_{W_k,\rho} \leq L_o\ \varepsilon|\rho|^{e_o}
\end{array}\right\} \leq L\ \varepsilon|\rho|^{e_o}$$

Moreover by (10.13) we have for $\nu \in \Delta_{e+1}$

$$\rho^\nu\|a_i^\nu\|^i_{U_i^5} \leq \gamma_\nu K_o\ \varepsilon|\rho|^{e_o}$$

$$\rho^\nu\|\varphi_{ij}^\nu\|^*_{U^5_{ij}} \leq \gamma_\nu K_o\ \varepsilon|\rho|^{e_o}\ .$$

By shrinking U_i^1, U_i^2, U_i^5 and intersecting with W_k^* we can assume that the two first and the two last norms are defined with respect to the chart W_k^* which only means multiplying L and K_o by a geometrical constant C>0. Now we can apply (8.1) to get: There is a constant K_1 depending only on K_o and the O-situation and there is a triangular condition $E_1 \subset E$ depending on L, E and the e_o-situation, such that for $(\varepsilon,\theta) \in E_1$ the following holds:

There are extensions σ_k^{e+1}, $T^{e+1}_{(k)i}$, $\check{T}^{e+1}_{(k)i}$ of σ_k^e, $T^e_{(k)i}$, $T^e_{(k)i}$ respectively defined over W_k, $W_k\cap U_i^6$ respectively, such that we have the commutative diagram

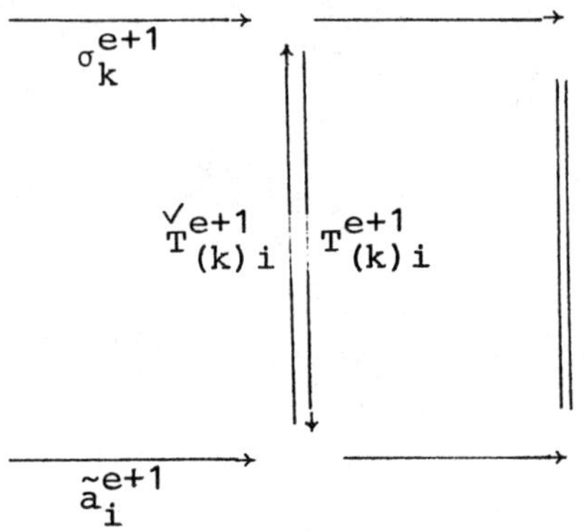

(Dia 5_{e+1})

and the estimates, $\nu \in \Delta_{e+1}$, independent of $e \geq e_o$:

$$\rho^{\nu}\|\sigma_k^{\nu}\|_{W_k}^{k} \leq \gamma_{\nu} \; K_1 \; \varepsilon|\rho|^{e_o}$$

$$\rho^{\nu}\|\check{T}^{\nu}_{(k)i}\|^{k}_{W_k \cap U_i^6} \leq \gamma_{\nu} \; K_1 \; \varepsilon|\rho|^{e_o} \qquad (8)$$

$$\rho^{\nu}\|T^{\nu}_{(k)i}\|_{W_k \cap U_i^6} \leq \gamma_{\nu} \; K_1 \; \varepsilon|\rho|^{e_o} ,$$

where σ_k^{ν}, $\check{T}^{\nu}_{(k)i}$, $T^{\nu}_{(k)i}$ are the coefficients of order e+1 of the extensions respectively.

(10.15) In order to extend the relating homomorphisms $\tau^{e}_{\mu k \ell}$ we consider on $W_{k\ell} \cap U^6_{ij}$ the series

$$\tilde{\tau}^{e+1}_{\mu k\ell,ij} = \mathrm{red}_{e+1} \; \check{T}^{e+1}_{(k)i} \; \tilde{\varphi}^{e+1}_{ij} \; T^{e+1}_{(\ell)j}$$

and $$\tilde{\tau}^{e}_{\mu k\ell,ij} = \tilde{\tau}^{e+1}_{\mu k\ell,ij}|S_e = \mathrm{red}_e \; \check{T}^{e}_{(k)i} \; \varphi^{e}_{ij} \; T^{e}_{(\ell)j} .$$

We are going to apply Lemma (8.3) to the following diagrams on $D_{\mu k\ell}$ resp. $D_{\mu k\ell} \cap U^6_{ij}$, where i = j is possible.

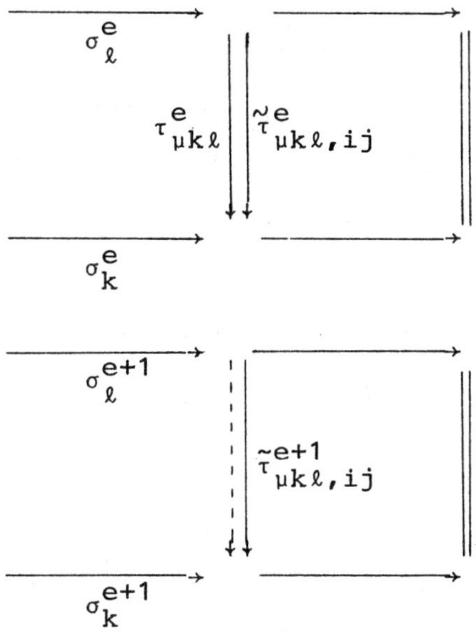

To do this we first have to give estimates for $\tilde{\tau}^e_{\mu k\ell,ij}$:

(10.15.1) Since $D_{\mu k\ell} \subset\subset W_\ell$ we can multiply L by a geometrical constant C to obtain

$$\| T^e_{(\ell)j} - T^{e_o}_{(\ell)j} \|^k_{D_{\mu k\ell} \cap U^7_{j,\rho}} \leq CL\ \varepsilon |\rho|^{e_o}$$

and similarly

$$\| \varphi^e_{ij} - \varphi^{e_o}_{ij} \|^k_{D_{\mu k\ell} \cap U^7_{ij,\rho}} \leq CL\ \varepsilon |\rho|^{e_o} .$$

Now using the product rule decomposition as in (5.8) we obtain a new constant L^* depending on L and the e_o-situation such that

$$\| \tilde{\tau}^e_{\mu k\ell,ij} - \tilde{\tau}^{e_o}_{\mu k\ell,ij} \|_{D_{\mu k\ell} \cap U^7_{ij,\rho}} \leq L^*\ \varepsilon |\rho|^{e_o}$$

if $(\varepsilon,\theta) \in E_2$, where $E_2 \subset E_1$ is a new triangular condition depending on the e_o-situation.

(10.15.2) If $\tau^{\nu}_{\mu k\ell,ij}$ for $\nu \in \Delta_{e+1}$ are the coefficients of $\tau^{e+1}_{\mu k\ell,ij}$ of order e+1 we obtain by calculating these coefficients from the definition of $\tau^{e+1}_{\mu k\ell,ij}$ and using estimates of type (5.8) for the e-terms, cf. (7.1), the following: If (ε,θ) satisfies still another triangular condition $E_3 \subset E_2$ depending on L, there is a constant K_2 depending only on K_o, K_1 and the geometrical situation such that

$$\rho^{\nu}\|\tau^{\nu}_{\mu k\ell,ij}\|_{D_{\mu k\ell}\cap U^7_{ij}} \leq \gamma_\nu \; K_2 \; \varepsilon|\rho|^{e_o} .$$

(10.15.3) Now we apply Lemma (8.3). There is a constant K_3 depending only on K_2 and the O-situation and there is a triangular condition $E_4 \subset E_3$ depending on L, E_3 and the e_o-situation such that for $(\varepsilon,\theta) \in E_4$ the following holds: There is an extension $\tau^{e+1}_{\mu k\ell}$ of $\tau^{e}_{\mu k\ell}$ on $D_{\mu k\ell}\times S_{e+1}$ such that we have the commutative diagram

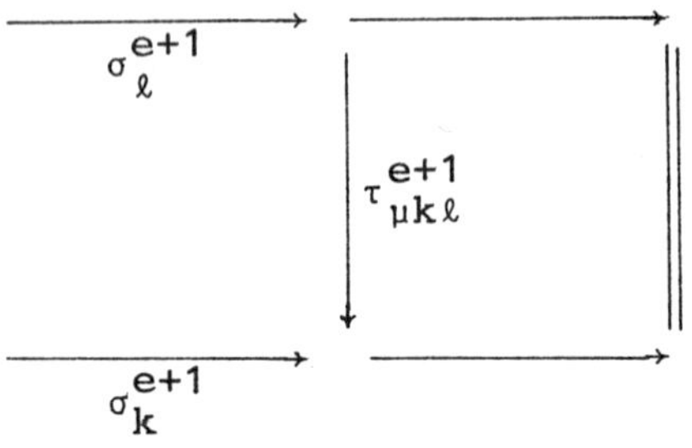

and the estimate

$$\rho^{\nu}\|\tau^{\nu}_{\mu k\ell}\|_{D_{\mu k\ell}} \leq \gamma_\nu \; K_3 \; \varepsilon|\rho|^{e_o} .$$

(10.16) Final choice of L_o, L, ε, θ.

We choose now L_o of (10.4.2) to be any constant $L_o > K_3$ and we choose L of (10.4.2) as in (10.11) such that L depends on L_o.

Note that $L_1, L_2, \ldots, L_4^*$ depend on L_o too. For (ε,θ) we assume that $(\varepsilon,\theta) \in E_4$. Then all the triangular conditions of § 10 are satisfied. By our procedure we have shown that all terms in (10.4.2) of the e-situation can be extended to the (e+1)-situation without changing the now defined constants of the estimates. This completes the construction of a deformation which for any finite order e is e-versal.

§ 11 Constructing morphisms of deformations.

A sequence of deformations $(\mathcal{F}_e, S_e)$, $e \in \mathbb{N}$, of $\mathcal{F}_o$ together with morphisms

$$\Phi_{e_1e_2} = (\varphi_{e_1e_2}, f_{e_1e_2}) : (\mathcal{F}_{e_2}, S_{e_2}) \to (\mathcal{F}_{e_1}, S_{e_1})$$

for $e_2 \geq e_1$ is called a <u>formal deformation</u> of $\mathcal{F}_o$ if there exists an ideal $\mathcal{J}$ in some formal power series ring $\mathbb{C}\{x_1,\dots,x_k\}$ such that

(i) each S_e is the germ of $\mathcal{J} + \mathfrak{m}^{e+1}/\mathfrak{m}^{e+1}$,

(ii) each $f_{e_1e_2}: S_{e_1} \longrightarrow S_{e_2}$ is the canonical imbedding,

(iii) $\Phi_{e_1e}\Phi_{ee_2} = \Phi_{e_1e_2}$ for $e_2 \geq e \geq e_1$, cf. [18], [21].

An analytic deformation $(\mathcal{F}, S)$ gives rise to a formal deformation by setting $\mathcal{F}_e = \mathcal{F}/\mathfrak{m}_S^{e+1}\mathcal{F}$. $(\mathcal{F}, S)$ is called <u>formally versal</u> etc. if the associated formal deformation is versal etc. in the category of <u>formal</u> deformations.

The formal deformation $\{(\mathcal{F}_e, S_e), \Phi_{de}\}$ is called <u>strongly formally complete</u> if for any other formal deformation $\{(\mathcal{F}'_e, S'_e), \Phi'_{de}\}$ and any morphism Θ as in the following diagram there exists a morphism $\tilde{\Theta}$ making the diagram commutative.

$$\begin{array}{ccc}
(\mathcal{F}_{e+1}, S_{e+1}) & \xrightarrow[\Phi_{e,e+1}]{} & (\mathcal{F}_e, S_e) \\
\downarrow \tilde{\Theta} & & \downarrow \Theta \\
(\mathcal{F}'_{e+1}, S'_{e+1}) & \xrightarrow[\Phi'_{e,e+1}]{} & (\mathcal{F}'_e, S'_e) .
\end{array}$$

If an analytic deformation $(\mathcal{F}, S)$ is formally versal, it is strongly formally complete since each $\mathcal{F}_e = \mathcal{F}/\mathfrak{m}_S^{e+1}\mathcal{F}$ is then a projective object in the category of deformations of order e, cf. [8], [21]. We are going to prove

<u>Theorem III:</u> <u>Let</u> X <u>be an arbitrary complex space and let</u> $\mathcal{F}_o$ <u>be a coherent analytic sheaf on</u> X <u>with compact support. If</u> $(\mathcal{F}, S)$ <u>is a deformation of</u> $\mathcal{F}_o$ <u>which is strongly formally complete (formally versal) it is already complete (versal).</u>

Especially the deformation $(\mathcal{F}, S)$ of § 10 is versal, by which theorem I is proved.

(11.1) Let $(\mathcal{F}, S)$ and $(\mathcal{G}, T)$ be analytic deformations of $\mathcal{F}_o$ and let $(\mathcal{F}, S)$ be strongly formally complete. Let $S \subset \mathbb{C}^m$ be defined by the holomorphic functions.

$$h_\kappa = \sum_{|\mu|>o} h_\kappa^\mu s^\mu \quad , \kappa = 1,\ldots,k$$

such that S_e is given by $h_\kappa^e = \sum_{o<|\mu|\leq e} h_\kappa^\mu s^\mu$ and let $\mathcal{F}_e = \mathcal{F}|S_e$.

Further we denote by $\mathfrak{J} \subset \mathbb{C}\langle t_1,\ldots,t_n\rangle$ the defining ideal of T and by $\mathfrak{J}_e = \mathfrak{J} + \mathfrak{m}^{e+1}/\mathfrak{m}^{e+1}$ its truncation of order e. There exists e_o such that $\Lambda(\mathfrak{J}_e) = \Lambda(\mathfrak{J}_{e_o}) = \Lambda$ for $e \geq e_o$. We denote by ω_λ, $\lambda \in \Lambda$, the GW-family of $\mathfrak{J}$ such that $\omega_\lambda^e = \omega_\lambda|\mathbb{C}_e\langle t\rangle$ give the GW-family of $\mathfrak{J}_e$. Let $\rho_o \in \mathbb{R}_+^m$ be such that

$$\|\omega_\lambda - t^\lambda\|_{\rho_o} \leq \frac{1}{2\operatorname{card}\Lambda} \rho_o^\lambda .$$

As in the previous sections we use the notations $\rho = \theta\rho_o$ for $0<\theta\leq 1$, and $\mathrm{red}^{e+1}f$ for the (e+1)-homogeneous part of

$\mathrm{red}_{e+1} f = \mathrm{red}_{\mathcal{J}_{e+1}} f$. Let $\mathcal{G}_e = \mathcal{G}|T_e$, where T_e is the germ of $\mathcal{J}_e$.

(11.2) We first choose a finite special covering, (9.1), $\mathfrak{U} = (\mathfrak{U}_i^*)_{i \in I}$ of Supp $\mathcal{F}_o$ with special charts $\mathfrak{U}_i^* = (U_i^*, Y_i, j_i, \Omega_i, G_i^*)$, $U_i^* = j_i^{-1} G_i^*$ and G_i cc Ω_i a privileged polydomain with respect to $\mathcal{O}_X$ and $\mathcal{F}_o$. Let $\mathfrak{U}''$ cc $\mathfrak{U}'$ cc $\mathfrak{U}$ cc $\mathfrak{U}^*$ be special coverings of Supp $\mathcal{F}_o$ obtained by shrinking such that the corresponding domains G_i'' cc G_i' cc G_i cc G_i^* are all privileged with respect to $\mathcal{O}_X$ and $\mathcal{F}_o$. We can assume that there is an index subset $I_o \subset I$ with :

a) if $i \in I \setminus I_o$ then $\overline{U}_i^* \cap \operatorname{Supp} \mathcal{F}_o = \emptyset$

b) Supp $\mathcal{F}_o \subset \bigcup_{i \in I_o} U_i''$

c) if $i \in I_o$ then $\overline{U}_i^* \subset \bigcup_{j \in I} U_j''$.

(11.3) Since $\mathcal{F}$ resp. $\mathcal{G}$ are given deformations of $\mathcal{F}_o$ defined over $X \times S$ resp. $X \times T$ we can find homomorphisms a_i resp. σ_{ji} defined over $U_i \times S$ resp. $U_{ij} \times S$ and b_i resp. τ_{ji} defined over $U_i \times T$ resp. $U_{ij} \times T$ representing $\mathcal{F}$ and $\mathcal{G}$ as described in the following diagrams

$$\begin{array}{ccccccc}
\mathcal{O}_{X\times S}^{q_i} & \xrightarrow{a_i} & \mathcal{O}_{X\times S}^{p_i} & \longrightarrow & \mathcal{F} & \longrightarrow & 0 \\
 & & \downarrow{\scriptstyle \sigma_{ji}} & & \| & & \\
\mathcal{O}_{X\times S}^{q_j} & \xrightarrow{a_j} & \mathcal{O}_{X\times S}^{p_i} & \longrightarrow & \mathcal{F} & \longrightarrow & 0
\end{array}$$

$$\begin{array}{ccccccc}
\mathcal{O}_{X\times T}^{q_i} & \xrightarrow{b_i} & \mathcal{O}_{X\times T}^{p_i} & \longrightarrow & \mathcal{G} & \longrightarrow & 0 \\
 & & \downarrow{\scriptstyle \tau_{ji}} & & \| & & \\
\mathcal{O}_{X\times T}^{q_j} & \xrightarrow{b_j} & \mathcal{O}_{X\times T}^{p_j} & \longrightarrow & \mathcal{G} & \longrightarrow & 0
\end{array}$$

These homomorphisms can be assumed to satisfy the following conditions.

(1) If we restrict the above diagrams to X we obtain the representation

$$\begin{array}{ccccccc} \mathcal{O}_X^{q_i} & \xrightarrow[a_i^o]{} & \mathcal{O}_X^{p_i} & \xrightarrow[p_i^o]{} & \mathcal{F}_o & \longrightarrow & 0 \\ & & \Big\downarrow \sigma_{ji}^o & & \Big\| & & \\ \mathcal{O}_X^{q_j} & \xrightarrow[a_j^o]{} & \mathcal{O}_X^{p_j} & \xrightarrow[p_j^o]{} & \mathcal{F}_o & \longrightarrow & 0 \end{array}$$

with $a_i^o = b_i^o$ and $\sigma_{ji}^o = \tau_{ji}^o$.

(2) The homomorphisms b_i, τ_{ji} are bounded and reduced with respect to the norms $\| \; \|_{U_i,\rho_o}$ and $\| \; \|^*_{U_{ij},\rho_o}$, and a_i, σ_{ji} have coefficients with finite norms $\| \; \|_{U_i}$, $\| \; \|^*_{U_{ij}}$ when they are represented by power series in $s_1,\dots,s_m$.

(3) If i is not in I_o then $a_i = 0$ and $b_i = 0$, and if i or j is not in I_o then $\sigma_{ji} = 0$ and $\tau_{ji} = 0$.

As before we use the notation $a_i^e = a_i | U_i \times S_e$ etc.

(11.4) We are going to construct m-tuples

$$f^e = (f_1^e,\dots,f_m^e) , \; f_\mu^e \in \mathbb{C}_e\langle t_1,\dots,t_n\rangle$$

and homomorphisms on $U_i \times T_e$

$$\Phi_i^e : \mathcal{O}_{X\times T_e}^{p_i} \longrightarrow \mathcal{O}_{X\times T_e}^{p_i}$$

$$\Psi_i^e : \mathcal{O}_{X\times T_e}^{q_i} \longrightarrow \mathcal{O}_{X\times T_e}^{q_i}$$

with the following properties.

(a) $\Phi_i^o = \mathrm{id}$, $\Psi_i^o = \mathrm{id}$, $f^o = 0$,

(b) f^e, Φ_i^e, Ψ_i^e extend f^{e-1}, Φ_i^{e-1}, Ψ_i^{e-1} and are bounded and reduced,

(c) $\Phi_i^e \to \Phi_i$, $\Psi_i^e \to \Psi_i$, $f^e \to f$ in the sense of Lemma (11.5),

(d) the following relations are satisfied.

(I_e) $\quad h_\kappa^e f^e \in \mathcal{J} + \mathfrak{m}^{e+1}$ $\quad$ for $\kappa = 1,\ldots,k$

(II_e) $\quad \Phi_i^e(a_i^e \circ f^e) = b_i^e \Psi_i^e$ $\quad$ on $U_i \times T_e$

(III_e) $\quad \Phi_j^e(\sigma_{ji}^e \circ f^e) = \tau_{ji}^e \Phi_i^e + b_j^e F_{ji}^e$ $\quad$ for some auxiliary homotopy homomorphism F_{ji}^e on $U_{ij} \times T_e$.

Then by (I) f defines a morphism $T \to S$, by (II) isomorphisms $\varphi_i : (\mathrm{id} \times f)^* \mathcal{F} \to \mathcal{G}$ over $U_i \times T$ are defined, and by (III) $\varphi_i = \varphi_j$ on $U_{ij} \times T$. By this a morphism $(\mathcal{F}, S) \to (\mathcal{G}, T)$ of the deformation is defined. Now the above matrices will be constructed by proving the following

<u>(11.5) Lemma:</u> <u>There exists a triangular condition</u> E <u>for</u> (ε,θ) <u>such that if</u> $(\varepsilon,\theta) \in E$ <u>the following holds.</u>

<u>If</u> f^e, Φ_i^e, Ψ_i^e, $e \geq e_o$, <u>exist with the relations</u> (I_e), (II_e), (III_e) <u>and if</u>

$$\|f^e\|_\rho \leq \varepsilon$$

$$\|\Phi_i^e - \Phi_i^o\|_{U_i,\rho} \leq \varepsilon$$

$$\|\Psi_i^e - \Psi_i^o\|_{U_i,\rho} \leq \varepsilon \ ,$$

<u>there exist extensions</u> f^{e+1}, Φ_i^{e+1}, Ψ_i^{e+1} <u>with relations</u> (I_{e+1}),

(II_{e+1}), (III_{e+1}) and with

$$\|f^{e+1}\|_\rho \leq \varepsilon$$

$$\|\Phi_i^{e+1} - \Phi_i^o\|_{U_i,\rho} \leq \varepsilon$$

$$\|\Psi_i^{e+1} - \Psi_i^o\|_{U_i,\rho} \leq \varepsilon .$$

The rest of § 11 is devoted to the proof of (11.5).

(11.6) We can represent h_κ, a_i, σ_{ji} by power series

$$h_\kappa = \sum_{o<|\mu|} h_\kappa^\mu s^\mu , \quad a_i = \sum_{o\leq|\mu|} a_i^\mu s^\mu , \quad \sigma_{ji} = \sum_{o\leq|\mu|} \sigma_{ji}^\mu s^\mu .$$

Since the coefficients can be assumed to be bounded over U_i resp. U_{ij}, we can choose $\varepsilon>0$ so small that

$$\sum_{1\leq|\mu|} \|a_i^\mu\|_{U_i} \varepsilon^{|\mu|-1} , \quad \sum_{1\leq|\mu|} \|\sigma_{ji}^\mu\|^*_{U_{ij}} \varepsilon^{|\mu|-1} \leq C_1$$

$$\sum_{2\leq|\mu|} \|a_i^\mu\|_{U_i} \varepsilon^{|\mu|-2} , \quad \sum_{2\leq|\mu|} \|\sigma_{ji}^\mu\|^*_{U_{ij}} \varepsilon^{|\mu|-2} \leq C_2$$

$$\sum_{2\leq|\mu|} |h_\kappa^\mu| \varepsilon^{|\mu|-2} \leq C_2$$

where C_1, $C_2 > 0$ are some constants.

Next we assume that $\overset{e}{f}{}^o$, $\overset{e}{\Phi}{}_i^o$, $\overset{e}{\Psi}{}_i^o$ have been constructed already with respect to $\mathfrak{U}^*$, which is possibly by algebraic considerations. Moreover after restricting $\varepsilon>0$ we choose $0<\theta<1$ so small that

$$\|b_i - b_i^o\|_{U_i,\rho} \leq \varepsilon \qquad \|\overset{e}{f}{}^o\|_\rho \leq \varepsilon$$

$$\|\tau_{ji}-\tau_{ji}^o\|^*_{U_{ij},\rho} \leq \varepsilon \qquad \|\overset{e}{\Phi}{}_i^o-\Phi_i^o\|_{U_i,\rho} \leq \varepsilon$$

$\|b_i - b_i^{e_o}\|_{U_i,\rho} \leq \varepsilon^2$ $\qquad \|\Psi_i^{e_o} - \Psi_i^o\|_{U_i,\rho} \leq \varepsilon$.

$\|\tau_{ji} - \tau_{ji}^{e_o}\|^*_{U_{ij},\rho} \leq \varepsilon^2$

This gives us the first triangular condition E_1 for (ε,θ) .

(11.7) Assume now that $(\varepsilon,\theta) \in E_1$ and that f^e, Φ_i^e, Ψ_i^e have been constructed with the estimates of (11.5), $e \geq e_o$. As in (11.4) they define a morphism $\Theta_e : (\mathcal{F}_e, S_e) \to (\mathcal{G}_e, T_e)$ of the deformations. Since $(\mathcal{F}, S)$ is supposed to be strongly formally complete, there exists an extension Θ_{e+1} of Θ_e . By flatness this implies the existence of extensions of f^e, Φ_i^e, Ψ_i^e as morphisms. But if we shrink to $\mathcal{U}'$ then the extensions can be assumed to be bounded and reduced of the form

$f^e + \Sigma f^\nu t^\nu$, $\phi_i^e + \Sigma\phi_i^\nu t^\nu$, $\psi_i^e + \Sigma\psi_i^\nu t^\nu$,

where summation is over $\nu \in \Delta_{e+1}$. Moreover we must have the relations mod $\mathcal{J} + \mathfrak{m}^{e+2}$:

$$h_\kappa \circ (f^e + \Sigma f^\nu t^\nu) \equiv 0$$

$$(\Phi_i^e + \Sigma\Phi_i^\nu t^\nu)(a_i^{e+1} \circ (f^e + \Sigma f^\nu t^\nu)) \equiv (b_i^e + \Sigma b_i^\nu t^\nu)(\Psi_i^e + \Sigma\Psi_i^\nu t^\nu)$$

$$(\Phi_j^e + \Sigma\Phi_j^\nu t^\nu)(\sigma_{ji}^{e+1} \circ (f^e + f^\nu t^\nu)) \equiv (\tau_{ji}^e + \Sigma\tau_{ji}^\nu t^\nu)(\Phi_i^e + \Sigma\Phi_i^\nu t^\nu)$$
$$+ (b_j^e + \Sigma b_j^\nu t^\nu)(F_{ji}^e + \Sigma F_{ji}^\nu t^\nu)$$

where F_{ji}^ν can be found algebraically to extend F_{ji}^e. In addition it is easy to see that also mod $\mathcal{J} + \mathfrak{m}^{e+2}$ we have

$$h_\kappa \circ (f^e + \Sigma f^\nu t^\nu) \equiv h_\kappa \circ f^e + \sum_{|\mu|=1} h_\kappa^\mu (\Sigma f^\nu t^\nu)^\mu$$
$$= h_\kappa \circ f^e + \sum_{\nu\in\Delta_{e+1}} t^\nu \sum_{|\mu|=1} h_\kappa^\mu (f^\nu)^\mu$$

and similarly for the other compositions. If we define the coefficients $h^{\nu}_{\kappa e}$, G^{ν}_{i}, H^{ν}_{ji} by

$$\sum_{\nu\in\Delta_{e+1}} h^{\nu}_{\kappa e} t^{\nu} = \mathrm{red}_{e+1}\ h_{\kappa}\circ f^{e}$$

$$\sum_{\nu\in\Delta_{e+1}} G^{\nu} t^{\nu} = \mathrm{red}_{e+1}\ (\Phi^{e}_{i}(a^{e}_{i}\circ f^{e}) - b^{e}_{i}\Psi^{e}_{i})$$

$$\sum_{\nu\in\Delta_{e+1}} H^{\nu}_{ji} t^{\nu} = \mathrm{red}_{e+1}\ (\Phi^{e}_{j}(\sigma^{e}_{ji}\ f^{e}) - \tau^{e}_{ji}\Phi^{e}_{i} - b^{e}_{j}F^{e}_{ji})$$

(note that by (5.4) there are no coefficients of order $\leq e$) we obtain the equations

$$(I_{*})\qquad \sum_{|\mu|=1} h^{\mu}_{\kappa}(f^{\nu})^{\mu} = -\,h^{\nu}_{\kappa e}$$

$$(II_{*})\qquad \Phi^{\nu}_{i}a^{o}_{i} + \sum_{|\mu|=1} a^{\mu}_{i}(f^{\nu})^{\mu} = b^{\nu}_{i} - a^{o}_{i}\Psi^{\nu}_{i} - G^{\nu}_{i}$$

$$(III_{*})\qquad \Phi^{\nu}_{j}\sigma^{o}_{ji} - \sigma^{o}_{ji}\Phi^{\nu}_{i} + \sum_{|\mu|=1} \sigma^{\mu}_{ji}(f^{\nu})^{\mu} = \tau^{\nu}_{ji} + a^{o}_{j}F^{\nu}_{ji} - H^{\nu}_{ji}\ .$$

Here we have used $F^{o}_{ji} = 0$, $\Phi^{o}_{i} = \mathrm{id}$, $\Psi^{o}_{i} = \mathrm{id}$ and $a^{o}_{i} = b^{o}_{i}$, $\sigma^{o}_{ji} = \tau^{o}_{ji}$.

Applying now the epimorphisms P^{o}_{i} and P^{o}_{j} we obtain the equations

$$(II')\qquad P^{o}_{j}\Phi^{\nu}_{i}a^{o}_{i} + \sum_{|\mu|=1} P^{o}_{i}a^{\mu}_{i}(f^{\nu})^{\mu} = P^{o}_{i}b^{\nu}_{i} - P^{o}_{i}G^{\nu}_{i}$$

$$(III')\qquad P^{o}_{j}\Phi^{\nu}_{j}\sigma^{o}_{ji} - P^{o}_{i}\Phi^{\nu}_{i} + \sum_{|\mu|=1} P^{o}_{j}\sigma^{\mu}_{ji}(f^{\nu})^{\mu} = P^{o}_{j}\tau^{\nu}_{ji} - P^{o}_{j}H^{\nu}_{ji}$$

(11.8) We now interprete (I_{*}), (II'), (III') as the solutions of an operator as follows. Set

$$C^{o}(\mathcal{U}',\ p.) = \prod_{i} \mathrm{Hom}(U^{!}_{i},\ \mathcal{O}^{p_i}_{X},\ \mathcal{F}_{o})$$

$$C^{o}(\mathcal{U}',\ q.) = \prod_{i} \mathrm{Hom}(U^{!}_{i},\ \mathcal{O}^{q_i}_{X},\ \mathcal{F}_{o})$$

$$C^1(\mathfrak{U}', p.) = \prod_{(i,j)} \mathrm{Hom}(U'_{ij}, \mathcal{O}_X^{p_i}, \mathcal{F}_o)$$

where the last product is over all <u>ordered</u> pairs (i,j) of the nerve, and define

$$C^o(\mathfrak{U}', p.) \oplus \mathbb{C}^m \xrightarrow{A} C^o(\mathfrak{U}', q.) \oplus C^1(\mathfrak{U}', p.) \oplus \mathbb{C}^k$$

by

$$(\zeta_i)_i \oplus x \longmapsto (\zeta_i a_i^o + \sum_{|\mu|=1} P_i^o a_i^\mu x^\mu)_i \oplus (\zeta_j \sigma_{ji}^o - \zeta_i + \sum_{|\mu|=1} P_j^o \sigma_{ji}^\mu x^\mu) \oplus (\sum_{|\mu|=1} h_\kappa^\mu x^\mu)_\kappa$$

where we write $\zeta_i^\nu = P_i^o \Phi_i^\nu$.

(11.8.1) <u>The operator</u> A <u>has closed range in the usual Fréchet-topology</u>.

Proof: We consider the following diagram

$$\begin{array}{ccccc} C^o(\mathfrak{U}', \mathcal{H}om(\mathcal{F}_o, \mathcal{F}_o)) & \xhookrightarrow{J^o} & C^o(\mathfrak{U}', p.) & \xrightarrow{a^*} & C^o(\mathfrak{U}', q.) \\ \downarrow \delta & & \downarrow \tilde{\delta} & & \\ C^1(\mathfrak{U}', \mathcal{H}om(\mathcal{F}_o, \mathcal{F}_o)) & \xhookrightarrow{J^1} & C^1(\mathfrak{U}', p.) & & \end{array} .$$

Here δ is the usual coboundary operator of the alternating cochains and $\tilde{\delta}$, J^o, J^1, a^* are defined as follows.

$$\tilde{\delta}(\zeta_i)_i = (\zeta_j \sigma_{ji}^o - \zeta_i) , \quad a^*(\zeta_i)_i = (\zeta_i a_i^o)_i$$

$$J^o(\varphi_i)_i = (\varphi_i P_i^o)_i$$

$$J^1(\varphi_{ji})_{ji} = (P_j^o \varphi_{ji})_{ji} .$$

Since A is only a finite-dimensional pertubation of $\tilde{\delta} \oplus a^*$, it is sufficient to prove that $\tilde{\delta} \oplus a^*$ has closed range. Now $H^1(\mathfrak{U}', \mathcal{H}om(\mathcal{F}_o, \mathcal{F}_o)) = H^1(X, \mathcal{H}om(\mathcal{F}_o, \mathcal{F}_o))$ is finite-dimensional, which implies that $\operatorname{Im}\delta$ is closed. Since U_i is Stein, also a^* has closed range. Since J^1 is a topological imbedding, we conclude that $J^1(\operatorname{Im}\delta)$ is closed. Now let

$$x \oplus y = (x_i)_i \oplus (y_{ji})_{ji}$$

be in the closure of $\operatorname{Im}(\tilde{\delta} \oplus a^*)$. Then there exists a sequence $\zeta^\lambda \in C^o(\mathcal{U}', p.)$ such that $a^*\zeta^\lambda$ tends to x and $\tilde{\delta}\zeta^\lambda$ tends to y. Since $\operatorname{Im} a^*$ is closed we find a convergent sequence $\tilde{\zeta}^\lambda \to \tilde{\zeta}$ such that $a^*\zeta^\lambda = a^*\tilde{\zeta}^\lambda$ and $a^*\tilde{\zeta} = x$. Then $\zeta^\lambda = \tilde{\zeta}^\lambda + J^o\varphi^\lambda$ for certain φ^λ and

$$J^1\delta\varphi^\lambda = \tilde{\delta}\zeta^\lambda - \tilde{\delta}\tilde{\zeta}^\lambda$$

converges to $y - \tilde{\delta}\tilde{\zeta}$. Since $J^1(\operatorname{Im}\delta)$ is closed, there is $\tilde{\varphi}$ such that

$$y = \tilde{\delta}\tilde{\zeta} + J^1\delta\tilde{\varphi} = \tilde{\delta}(\tilde{\zeta} + J^o\tilde{\varphi}) .$$

Since also $x = a^*(\tilde{\zeta} + J^o\tilde{\varphi})$ we have shown that $x \oplus y \in \operatorname{Im}(\tilde{\delta} \oplus a^*)$.

(11.8.2) Now by the open mapping theorem and the definition of the natural Fréchet-topologies A has a pullback constant K_o with respect to $\mathcal{U}'' \subset\subset \mathcal{U}'$. Since by (I_*), (II'), (III')

$$(P_i^o b_i^\nu - P_i^o G_i^\nu)_i \oplus (P_j^o \tau_{ji}^\nu - P_j^o H_{ji}^\nu)_{ij} \oplus (-h_{\kappa e}^\nu)_\kappa$$

are in the image of A for $\nu \in \Delta_{e+1}$, we find solutions f^ν and ζ_i^ν on U_i'' such that

$$\|\zeta_i^\nu\|_{U_i''} , \|f^\nu\| \le K_o M$$

if we have the estimates

$$\|b_i^\nu\|_{U_i} , \|G_i^\nu\|_{U_i} , \|\tau_{ji}^\nu\|^*_{U_{ij}} , \|H_{ji}^\nu\|^*_{U'_{ij}} , |h_{\kappa e}^\nu| \le M .$$

(11.9) To obtain the final estimates in Lemma (11.5) we first have to give estimates of $h^{\nu}_{\kappa e}$, G^{ν}_{i} and H^{ν}_{ji} in (11.7). If we assume the estimates in Lemma (11.5) given, we find by a product decomposition as in § 5 estimates

$$\|\mathrm{red}^{e+1}\ b^{e}_{i}\Psi^{e}_{i}\|_{U_i,\rho} \leq 2\varepsilon^2$$

$$\|\mathrm{red}^{e+1}\tau^{e}_{ji}\Phi^{e}_{i}\|^{*}_{U_{ij},\rho} \leq 2\varepsilon^2 .$$

To estimate terms like $\Phi^{e}_{i}(a^{e}_{i}\circ f^{e})$ we can write this term as

$$a^{o}_{i} + \sum_{|\mu|=1} a^{\mu}_{i}(f^{e})^{\mu} + (\Phi^{e}_{i}-1)a^{o}_{i} + \sum_{2\leq|\mu|} a^{\mu}_{i}(f^{e})^{\mu} + (\Phi^{e}_{i}-1)\sum_{1\leq|\mu|} a^{\mu}_{i}(f^{e})^{\mu}$$

and by using $\|f^{e}\|_{\rho} \leq \varepsilon$ we obtain from this for the terms of order e+1

$$\|\mathrm{red}^{\,e+1}\Phi^{e}_{i}(a^{e}_{i}\circ f^{e})\|_{U_i,\rho} \leq 2\varepsilon^2 C_2 + 2\varepsilon\cdot\varepsilon C_1 = 2\varepsilon^2(C_1 + C_2) .$$

Similarly we get the estimates

$$\|\mathrm{red}^{e+1}\Phi^{e}_{j}(\sigma^{e}_{ji}\circ f^{e})\|^{*}_{U_{ij},\rho} \leq 2\varepsilon^2(C_1 + C_2)$$

$$|\mathrm{red}^{e+1}h_{\kappa}\circ f^{e}|_{\rho} \leq 2\varepsilon^2 C_2 .$$

(11.9.2) Now we construct a solution F^{e}_{ji} over $U'_{ij}\times T_e$ of the equation

$$b^{e}_{j}F^{e}_{ji} = \tau^{e}_{ji}\Phi^{e}_{i} - \Phi^{e}_{j}(\sigma^{e}_{ji}\circ f^{e})$$

by using a simplified version of the main Lemma (7.4) with $e_o = 0$. Since the right hand side is estimated by $2\varepsilon^2(1+C_1+C_2)$ we can find a triangular condition $E_2 \subset E_1$ such that for $(\varepsilon,\theta) \in E_2$ a solution F^{e}_{ji} of the above equation can be found satisfying

$$\|F^e_{ji}\|^*_{U'_{ij},\rho} \leq \varepsilon L_1$$

independent of $e \geq e_o$, where L_1 is a constant depending only on C_1, C_2 and the o-situation. Then by the definition of $h^\nu_{\kappa e}$, G^ν_i, H^ν_{ji} in (11.7) we obtain the estimates:

$$\rho^\nu |h^\nu_{\kappa e}| \quad , \quad \rho^\nu \|G^\nu_i\|_{U_i} \quad , \quad \rho^\nu \|H^\nu_{ji}\|^*_{U'_{ij}} \leq \gamma_\nu \; \varepsilon^2 L_2$$

where $L_2 \geqslant 1$ is a new constant depending only on L_1 by an algebraic expression. Moreover by the choice of $(\varepsilon,\theta) \in E_1$ we have

$$\rho^\nu \|a^\nu_i\| \quad , \quad \rho^\nu \|\tau^\nu_{ji}\|^*_{U_{ij}} \leq \gamma_\nu \; \varepsilon^2 \leq \gamma_\nu \; \varepsilon^2 L_2 \quad .$$

(11.10) By (11.8.2) and the estimates in (11.9) we find solutions ζ^ν_i and f^ν of the operator A satisfying

$$\zeta^\nu_i a^o_i + \sum_{|\mu|=1} P^o_i a^\mu_i (f^\nu)^\mu = P^o_i b^\nu_i - P^o_i G^\nu_i \qquad (\mathrm{II}')$$

$$\zeta^\nu_j \sigma^o_{ji} - \zeta^\nu_i + \sum_{|\mu|=1} P^o_j \sigma^\mu_{ji} (f^\nu)^\mu = P^o_j \tau^\nu_{ji} - P^o_j H^\nu_{ji} \qquad (\mathrm{III}')$$

$$\sum_{|\mu|=1} h^\mu_\kappa (f^\nu)^\mu = - h^\nu_{\kappa e} \qquad (\mathrm{I}_*)$$

and the estimates

$$\rho^\nu \|\zeta^\nu_i\|_{U''_i} \quad , \quad \rho^\nu \|f^\nu\| \leq \gamma_\nu \; \varepsilon^2 K_o L_2 \quad .$$

(11.11) Now since the norms of $\mathcal{F}_o$ are defined via the projections P^o_i we can find homomorphisms $\tilde{\Phi}^\nu_i$ such that $P^o_i \tilde{\Phi}^\nu_i = \zeta^\nu_i$ and such that

$$\rho^\nu \|\tilde{\Phi}^\nu_i\|_{U''_i} \leq \gamma_\nu \; 2\varepsilon^2 K_o L_2 \; .$$

Moreover by privilegedness and (II') we can find $\tilde{\Psi}_i^\nu$ on U_i'' satisfying

$$(II_*) \quad \tilde{\Phi}_i^\nu a_i^o + \sum_{|\mu|=1} a_i^\mu (f^\nu)^\mu = b_i^\nu - a_i^o \tilde{\Psi}_i^\nu - G_i^\nu$$

and $\quad \rho^\nu \| \tilde{\Psi}_i^\nu \|_{U_i''} \leq \gamma_\nu \varepsilon^2 L_3$,

where L_3 only depends on L_2 and the geometric situation. By (III') we can find $\tilde{F}_{ji}^\nu$, not necessarily with bounds, such that

$$(III_*) \quad \tilde{\Phi}_j^\nu \sigma_{ji}^o - \sigma_{ji}^o \tilde{\Phi}_i^\nu + \sum_{|\mu|=1} \sigma_{ji}^\mu (f^\nu)^\mu = \tau_{ji}^\nu + a_j^o \tilde{F}_{ji}^\nu - H_{ji}^\nu$$

is satisfied. By the now satisfied equations (I_*), (II_*), (III_*) we obtain with

$$f^{e+1} = f^e + \Sigma f^\nu t^\nu$$

$$\tilde{\phi}_i^{e+1} = \phi_i^e + \Sigma \tilde{\phi}_i^\nu t^\nu \qquad \text{on } U_i'' \times T_{e+1}$$

$$\tilde{\psi}_i^{e+1} = \psi_i^e + \Sigma \tilde{\psi}_i^\nu t^\nu \qquad \text{on } U_i'' \times T_{e+1}$$

extensions of f^e, ϕ_i^e, ψ_i^e with bounds satisfying (I_{e+1}), (II_{e+1}), (III_{e+1}) . In order to obtain extensions over $U_i \times T_e$ with bounds we have to make the following considerations.

(11.12) By the relations (I_{e+1}), (II_{e+1}), (III_{e+1}) the homomorphisms $\tilde{\phi}_i^{e+1}$ induce isomorphisms

$$\varphi_i^{e+1} : (id \times f^{e+1})^* \mathcal{F}_{e+1} \longrightarrow \mathcal{G}_{e+1} \qquad \text{over } U_i'' \times T_{e+1}$$

and $\varphi_i^{e+1} = \varphi_j^{e+1}$ over $U_{ij}'' \times T_{e+1}$. Hence a global isomorphism $\varphi^{e+1} : (id \times f^{e+1})^* \mathcal{F}_{e+1} \to \mathcal{G}_{e+1}$ is defined over $\bigcup_{i \in I} U_i'' \times T_{e+1}$ which can be trivially extended to $X \times T_{e+1}$. We are going to define Φ_i^{e+1}, Ψ_i^{e+1} on $U_i \times T_{e+1}$ extending Φ_i^e, Ψ_i^e.

If $i \in I \setminus I_o$ we can choose $\Phi_i^{e+1} = 0$, $\Psi_i^{e+1} = 0$. If $i \in I_o$ we have $U_i \subset \bigcup_{j \in I} U_j''$. Then we consider the relating homomorphisms

$$\tau_{ij}^{e+1} \tilde{\phi}_j^{e+1} (\sigma_{ji}^{e+1} \circ f^{e+1}) \qquad \text{on } U_i \cap U_j'' \times T_{e+1}$$

which induce the same isomorphism φ^{e+1} for the cokernels. Now we use amalgamation I, (4.1), and a simplified version of (8.3) to find a constant L_4 depending only on L_3 and the O-situation together with a triangular condition $E_3 \subset E_2$ such that for $(\varepsilon,\theta) \in E_3$ we can find an extension ϕ_i^{e+1} of ϕ_i^e on $U_i \times T_e$ such that ϕ_i^{e+1} induces φ^{e+1} for the cokernels and satisfies

$$\| \Phi_i^{e+1} - \Phi_i^e \|_{U_i,\rho} \leq L_4 \, \varepsilon^2$$

independent of $e \geq e_o$. Again by a simplified version of (7.2) we find a constant L_5 depending only on L_3, L_4 and the O-situation together with a triangular condition $E_4 \subset E_3$ such that for $(\varepsilon,\theta) \in E_4$ we can find an extension Ψ_i^{e+1} of Ψ_i^e over $U_i \times T_{e+1}$ satisfying

$$(II_{e+1}) \qquad b_i^{e+1} \Psi_i^{e+1} = \Phi_i^{e+1} (a_i^{e+1} \circ f^{e+1}) \qquad \text{on } U_i \times T_{e+1}$$

and

$$\| \Psi_i^{e+1} - \Psi_i^e \|_{U_i,\rho} \leq L_5 \, \varepsilon^2 .$$

Now since Φ_i^{e+1}, Φ_j^{e+1} induce the same isomorphism on $U_{ij} \times T_{e+1}$ we can find F_{ij}^{e+1} without bounds, such that (III_{e+1}) is verified.

To complete the proof of Lemma (11.5) we choose now $\varepsilon > 0$ in addition to $(\varepsilon,\theta) \in E_5$ so small that $\varepsilon K_o L_2 \leq 1$, $\varepsilon L_4 \leqq 1$, $\varepsilon L_5 \leq 1$. Then we have the estimates

$$\|f^{e+1}\|_\rho \leq \varepsilon$$

$$\|\phi_i^{e+1} - \phi_i^o\|_{U_i,\rho} \leq \varepsilon$$

$$\|\psi_i^{e+1} - \psi_i^o\|_{U_i,\rho} \leq \varepsilon$$

By this Lemma (11.5) and theorem III are proved.

Department of Mathematics
Stanford University
Stanford, California 94305

Fachbereich Mathematik
Universität Kaiserslautern
D-6750 Kaiserslautern
F.R. Germany

References

[1] Commichau, M.: Deformation kompakter komplexer Mannigfaligkeiten, Math. Ann. 213, 43-96 (1975)

[2] Donin, I. F.: Complete families of deformations of germs of complex spaces, Math. Sbornik 18, 397-406 (1972)

[3] Douady, A.: Le problème des modules pour les sous-espaces analytiques compacts d'une espace analytique donné, Ann. Inst. Fourier 16, 1-95 (1966)

[4] Douady, A.: Flatness and Privilege, Monographie No. 17, l'Enseignement Mathématique, Genève, 47-74 (1968)

[5] Douady, A.: Le problème des modules locaux pour les espaces $\mathbb{C}$-analytiques compacts, Ann. Sci. Ec. Norm. Sup. 7, 569-602 (1974)

[6] Forster, O. - Knorr, K.: Über die Deformationen von Vektorraumbündeln auf kompakten komplexen Räumen, Math. Ann. 209, 291-346 (1974)

[7] Forster, O. - Knorr, K.: Ein neuer Beweis des Satzes von Kodaira - Nirenberg - Spencer, Math. Z. 139, 257-291 (1974)

[7a] Forster, O. - Knorr, K.: Konstruktion verseller Deformationen kompakter komplexer Räume, Lecture Notes in Mathematics 705, Springerverlag 1979

[8] Frisch, J.: Aplatissement en Géometrie analytique, Ann. Sci. Ec. Norm. Sup. 1, 305-312 (1968)

[9] Galligo, A.: A propos du théorème de preparation de Weierstrass. Thèse, Seminaire Norguet, Lecture Notes in Mathematics 409, 543-579, Springerverlag 1974

[10] Galligo, A. - Houzel, C.: Module des singularités isolées d'après Verdier et Grauert, Astérisque 7,8, Soc. Math. France, 139 - 163 (1973)

[11] Grauert, H.: Über die Deformationen isolierter Singularitäten analytischer Mengen, Invent. Math. 15, 171 - 198 (1972)

[12] Grauert, H.: Der Satz von Kuranishi für kompakte komplexe Räume, Invent. Math. 25, 107 - 142 (1974)

[13] Hörmander, L.: An Introduction to complex analysis in several variables, van Nostraud 1966

[14] Hörmander, L.: L^2-estimates and existence theorems for the $\bar{\partial}$-operator. Acta Math. 113, 89 - 152 (1965)

[15] Pourcin, G.: Polycylindres privilegies et deformation de singularités isolées, Astérisque 16, Soc. Math. France, 145 - 173 (1974)

[16] Schlessinger, M.: Functors of Artin rings, Trans. Amer. Math. Soc. 130, 208 - 222 (1968)

[17] Schuster, H. W.: Infinitesimale Erweiterungen komplexer Räume, Comm. Math. Helvetici 45, 265 - 286 (1970)

[18] Schuster, H. W.: Formale Deformationstheorien, Habilitationsschrift München 1972

[19] Siu, Y. T.: Characterization of privileged polydomains, Trans. Amer. Math. Soc. 193, 329 - 357 (1974)

[20] Tiourina, G. N.: Locally semi-universal flat deformations of isolated singularities of complex spaces. Mathematics USSR-Izvestija 3, 967 - 999 (1969)

[21] Trautmann, G.: Deformation von isolierten Singularitäten kohärenter analytischer Garben, Math. Ann. 223, 71-89 (1976)

[22] Trautmann, G.: Deformations and moduli of coherent analytic sheaves with finite singularities, Seminaire Norguet 1957/77, 233-302, Lecture Notes in Mathematics 670, Springerverlag 1978.

ABCDEFGHIJ-AMS-89876543210